COUNTRY LAND & ITS USES

howard orem & suzen

NATUREGRAPH PUBLISHERS
Healdsburg   1974

Library of Congress Cataloging in Publication Data CIP

Orem, Howard, 1914–
    Country land & its uses.

    Includes index.
    1. Farmhouses. 2. Water-supply, Rural. 3. Organic gardening. 1. Land. 2. Water resources development. 3. Organic gardening. 4. Country life I. Suzen, joint author. II. Title.

TH4920.07   630   74-22154

Copyright © 1975 by Howard Orem and Suzen Snyder

ISBN 0-87961-031-X **Cloth Edition**
ISBN 0-87961-030-1 **Paper Edition**

Naturegraph Publishers Inc., Healdsburg, California 95448.

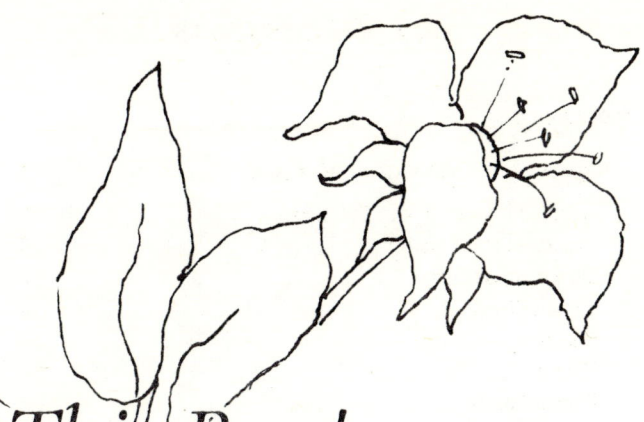

*This Book
is dedicated to
All Beneficial Life*

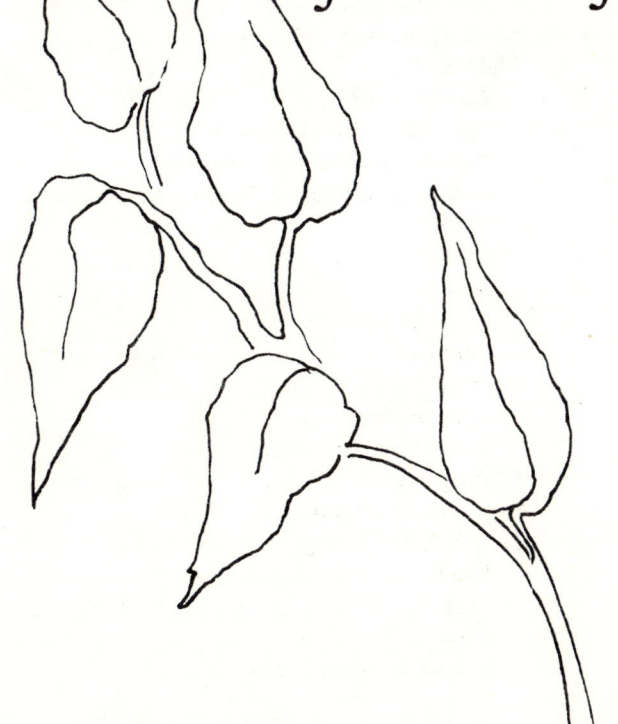

TABLE OF CONTENTS

Foreword. 7

Chapter I — Selection of Land   9

   Location Away From Pollution 11
   Availability of Water 13
   Direction of Slope 41
   Type of Soil 45
   Roads 55
   Climate 56
   Microclimate 62
   Vegetation 64
   Compatability of Neighbors 66
   Legal Aspects 67
   Meditation 71

Chapter II — Development of Land   73

   Roads 74
   Choosing a Housesite 79
   Clearing the Land 84
   Leveling Hills and Terracing 85
   Building Houses 87
      Selection of Materials 87
      Foundation With Basement 88
      Foundations Without Basement 95
      House Construction on Basement Foundation   97
      Floor 97
      Frame 98
      Roof 100
      Covering the Frame 102
      Windows 102
      Doors 104
      Outdoor Privy 105
      Septic Tank and Drainfield 106
   Building Fences 111
   Microclimate Development 113
   Using Nature's Energy 116
      Water 117
      Wind 121
      Sun 126
      Woodstove Water Heater 130

Woodstove	131
How to Start a Fire	132
Preventing Erosion	133
Plants Which Prevent Erosion	133
Stopping Streambank Erosion	134
Damming Erosion Gulleys to Make Ponds	136
How to Prevent a Fire	137

Chapter III — Development of Water   139

The Source	139
Constructing a Water Tank	141
Concrete	143
Wood	146
Other	146
Developing a Spring	146
Developing a Creek	150
Ponds	154
Rainwater	159
Developing a Watershed	160
Water Recharge Reservoirs	163
Dams	165
Earth	165
Concrete	167
Pumps	168
How to Buy	169
How to Install	169

Chapter IV — Organic Gardening   173

Choosing Location of the Garden	173
The Fence	175
Irrigation Water System	176
Compost	177
Preparing Soil Texture	181
Testing the Soil	183
Improving Soil Fertility	185
Planning the Garden	187
Companion Planting	188
Planting by the Moon	189
Planting the Seeds	191
Transplanting	193
Growth Chart	197
The Vegetables	199

Watering the Garden	250
Mulching	251
Fertilizing	254
Orem's Compost	256
Deficiency Symptoms	257
Fertilizers	261
Seasonal Protection	264
Weeds	271
Crop Rotation	272
Garden Pest Control	273
The Orchard	281
Fruit Trees	288
Berries and Grapes	293
The Herb Garden	297
Cold Frames and Greenhouses	299
In Conclusion	303
Index	304

FOREWORD

The purpose of this book is to aid the many people who are returning to the land in the selection and development of a self-sufficient environment. Many of the problems in the world today are the result of an impure and unbalanced system of dealing with the flow of life. We cannot continue to take energy from the earth without giving back something in return. We hope, through this book, to give you a feeling for the balance, and a desire to plant the seeds of a new life and a new world. A vast storehouse of secrets, discovered through years of research at Orem's Organic Gardens, is revealed in these pages. Most of the information within this book has been gathered from experience, which is still the best teacher. We encourage you to experience for yourself the many ways of nature.

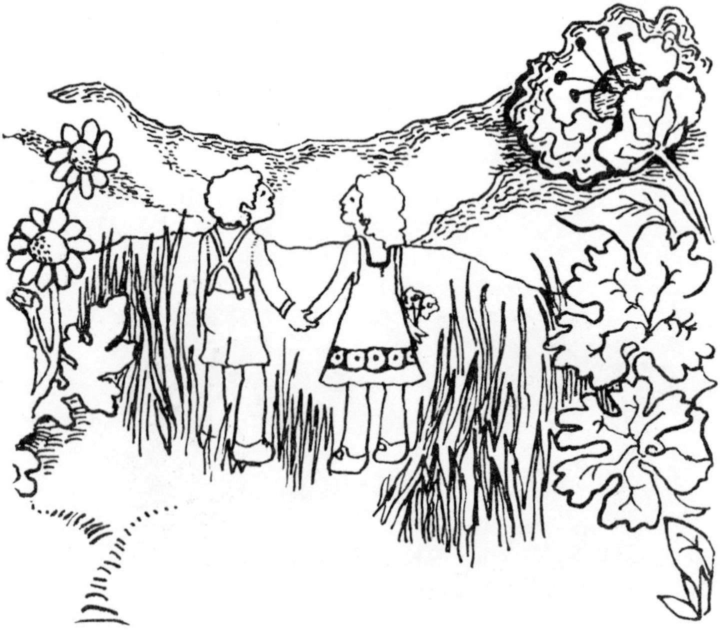

CHAPTER I
SELECTION OF LAND

Careful selection of land is very important because the success or failure of a farmstead, as in all phases of life, depends upon what you have to work with in the beginning. It is important to take your time and not be rushed into making a decision. Know what you want and be conscious of what you will need.

One universal rule to follow is that QUALITY is more important than QUANTITY. When buying land, it is much better to buy a small piece of usable land than a large piece of land which is not level enough, accessible enough, or fertile enough to be usable, even though this larger piece may be very inexpensive. These "bargains" usually turn out to be "bad deals", especially if you consider calculating the cost per acre in terms of usable acreage.

Spend as much time there as possible so as to see the land in its various cycles of natural change. This is especially important in seasonal areas, where there is a great variation in wet and dry conditions and hot and cold temperatures. It is good to familiarize yourself with the general area first, by moving or camping there before purchasing land in that area. Many people are enchanted immediately by the scenic beauty of an area and ignore the practical aspects of developing the land. Careful consideration and patience are essential in selecting a farmstead with the greatest potential for harmonious development.

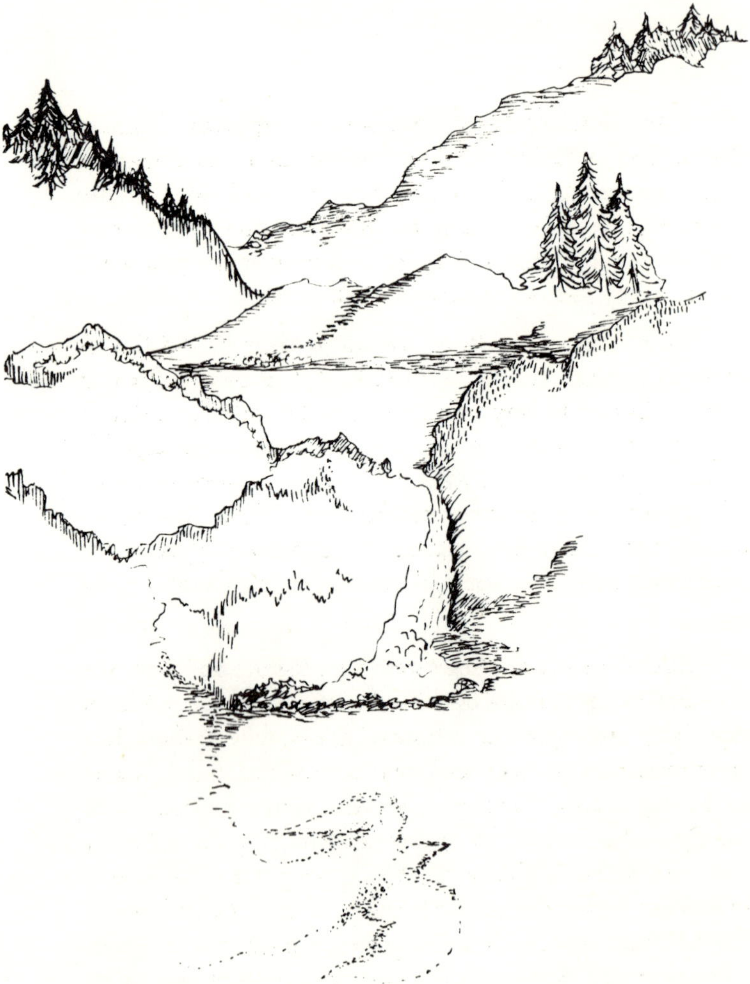

LOCATION AWAY FROM POLLUTION

Purity of Spirit begins with the purification of the body. Through ignorance, man has come close to destroying the earth. Only through awareness and harmony with nature can he hope to restore it. By keeping our bodies and our environments pure, our thoughts, feelings and energies positive, we can help to bring about a more creative, more conscious world.

Purification of the earth begins with each individual. Through the avoidance of pollutants and polluted areas, we maintain and develop the individual clarity necessary to bring about change. If we refuse to subsidize the industries that are destroying our earth and people, these industries will either vanish or be forced to use their imaginations in coming up with alternative natural products.

Meanwhile, we must all do what we can to handle all phases of life as naturally and as harmoniously as possible.

It is best to avoid locating a farmstead near major cities and industries for obvious reasons. If you must be near industrial centers, try to locate upwind and upstream from them so as to avoid pollutants carried by wind and water. Growing with or near chemicals and near pollution centers is harmful because soil containing pesticides and other chemicals soon becomes devoid of certain natural bacteria needed by plants to maintain healthy growth.

Never buy land used previously for commercial farming, or land located near commercial farms. Commercial farming land is usually saturated with pesticides, which may also be carried a considerable distance by wind and water. It is best to be as far away as possible

POISO
x x x
x x
x
may
kill all
livingthings
on
con
tact
for
ever
DDT
O
X
0
poison

from these commercial farms, as these chemicals also pollute subsurface water.

It is best to be as far away as possible from highways and railways and airports. Aside from air and noise pollution, the chemicals used in the manufacture of asphalt and for treating railways are harmful to plants and people. These pollutants may be carried by rain water as it flows over polluted lands on its way to streams and rivers. It is possible to overcome the hazards of roadway pollutants washing onto land by building a levee along the road with a sloping ditch between road and levee to allow water to run off.

Avoid land and watersheds near utility poles, as utility companies sometimes spray chemicals around the bases of poles to kill vegetation. Poles are also treated with creosote and other chemicals which are harmful. These chemicals may very easily be carried by wind and water, polluting air, springs, and even wells.

Keep in mind that water that comes from the ground is first deposited as rain on a watershed. Any water-soluble matter on or under the watershed can be dissolved in the water and carried to the underground reservoir. This is why so many wells and springs produce polluted water. It is a gross misconception to believe that all spring water is pure.

We can't undo what has already been done, but perhaps we can learn from past actions. It has become obvious that the chemical solutions to the various problems have created a much greater problem, that of imbalance. A new way will bring us a new world.

AVAILABILITY OF WATER

Water is perhaps the most important consideration

when searching for land. It is the Source of Life, but too much of it in the wrong places at the wrong times may also bring destruction. Through observation and careful control of water flows, it is possible to create a balance. If you know what to look for and how to work with what you find, maintaining the balance will be simple.

Look for land during the dry season when choosing in upland areas, as lack of water will be most evident at this time. Upland areas are lands high above a river or creek and are generally developed from the parent rock. Bottomland areas, which are low lands near rivers or creeks, should be chosen during the wet time of year, as flooding may be a problem. It is important to be sure that possible building sites will not be in the midst of seasonal flood waters. Indication of this is heavy silt deposits, as well as debris and brush caught on rocks, tree trunks and low limbs.

It is best to have a stream originating (headwaters) on your property. It is also desirable to have as many creeks and springs on the property as possible. Springs are generally a good source of water as they originate underground. The underlying conditions of the earth and rock formations direct the water to the surface. A good spring runs year 'round and would probably provide plentiful water if developed properly. A smaller spring, which may return underground during the dry season, may be a bit more difficult to locate and develop, but may still provide enough water for basic needs. It is wise, in either case, to have an alternate source of water and to store as much water as possible during the rainy season in tanks, ponds, or other reservoirs. It is of great advantage to find a water source high on the land, above future building sites. This will create gravity-flow water to your house and garden and will eliminate the hassle

SUBSURFACE WATER COLLECTION

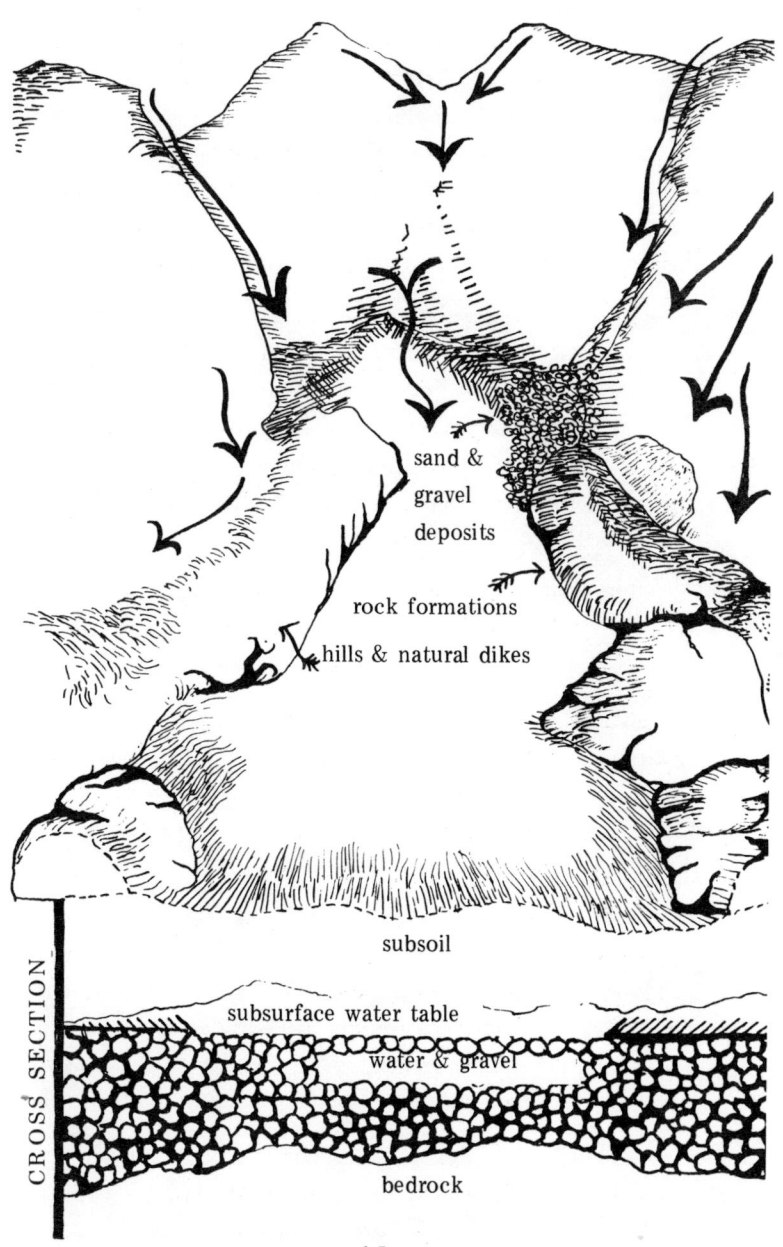

sand & gravel deposits

rock formations

hills & natural dikes

subsoil

subsurface water table

water & gravel

bedrock

CROSS SECTION

of pumping or hauling water. If there is a creek flowing rapidly with a fairly steep drop in elevation, there is a good chance of developing hydroelectric power there: FREE ELECTRICITY!

If there is not much sign of surface water such as springs, rivers or creeks, there may still be the possibility of developing underground springs or other subsurface water, especially in areas where there are seasonal heavy rains. Subsurface water may be developed by digging out a spring and inserting a springbox, drilling a well, digging a pond, or recessing other subsurface reservoirs.

The basic structure of the terrain will tell much about possibilities of subsurface water. Long, flat valleys with hills sloping down on two or more sides, forming a V, generally contain much subsurface water, especially if the soil contains much sand, silt, and gravel. Short, steep valleys may have some subsurface water, but chances of finding any are not as good.

Layered rock formations that slope toward a hill, dike or fault often catch much water and form subsurface as well as above-surface water flows.

Water is often collected behind natural dikes, such as large rocks, long low hills, and fallen trees, especially where the terrain slopes downward toward the dike. These dikes form natural dams and trap water behind them.

Outcroppings of porous rock on hills are often a sign of subsurface water, as are large rock-covered areas and large flat areas of land, especially where soil is porous. They form natural watersheds, as they catch much rainwater and take it below the surface rather than allowing it to run off down the hill.

If land shows very heavy natural water erosion, it is a sign that rain waters are running off and not collecting below the surface. This is usually caused by tight, nonporous soil that does not allow water to seep in. On the other hand, heavy limestone areas generally allow most rain waters to seep in very deeply and run into underground channels. Usually this water is too deep beneath the surface to develop. The same is true of areas with a deep layer of extremely porous topsoil. If soil is dry at bedrock, the chances of finding subsurface water are very small.

Where trees and brush are scrubby and grasses die early in the summer, there is probably a layer of bedrock just below the surface and a very shallow layer of topsoil. The possibilities of subsurface water here are very slim, as rain waters seep in very shallowly and evaporate quickly, preventing water from collecting below the surface.

Vegetation tells much about the availability of water below the surface. Just the fact that something can grow there at all indicates the presence of some water. The larger, the healthier, the greener, and the more abundant the vegetation, the better the chance of finding subsurface water. Some plants need a continuous supply of water to survive. Those plants are an almost certain indication of subsurface water.

Here are a few:

adder's tongue	marsh fern
ash tree	poison sumac
azalea, wild	rattlesnake fern
cabbage palmetto	skunk cabbage
cattail	spignate
crested woodfern	water pennywort
hedge hyssop	willow tree
horsetail	woodwardia fern

PLANT 6-12 inches high

ADDER'S TONGUE

20-50 feet

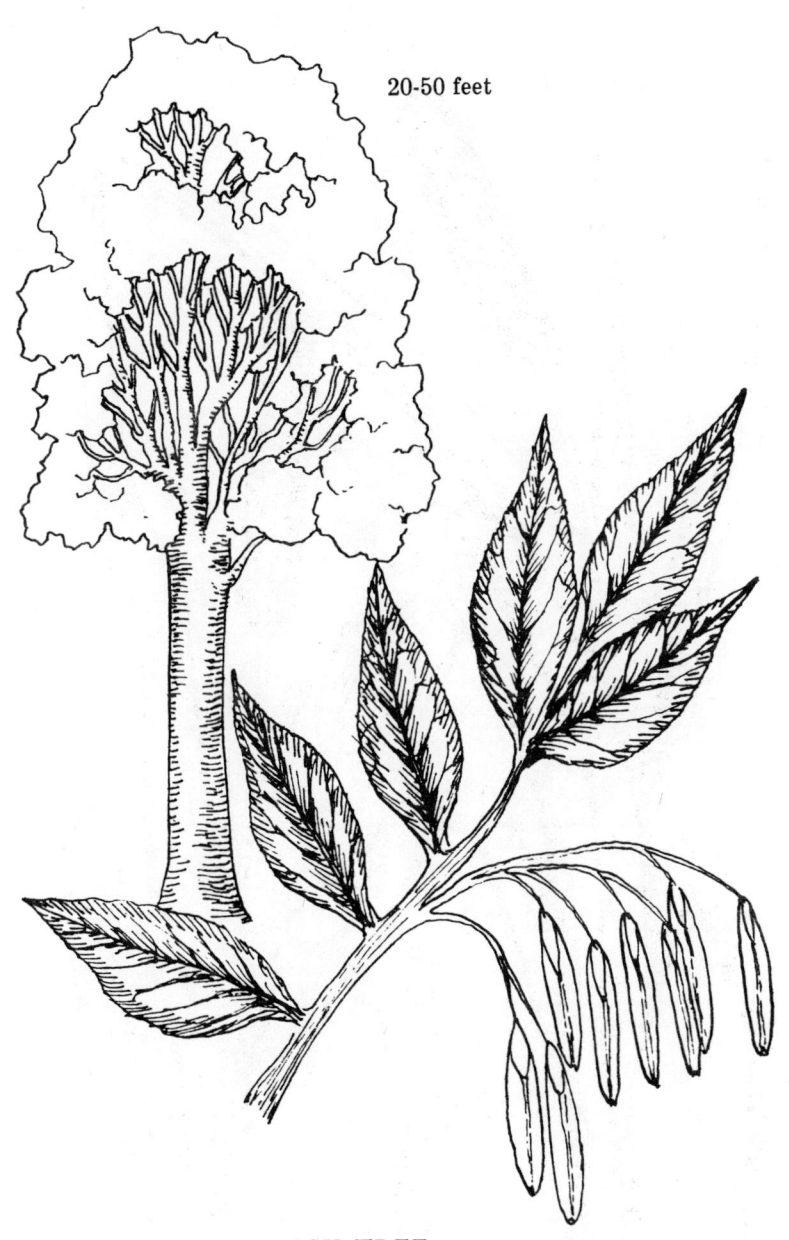

ASH TREE

CATTAIL 4-8 feet

SPORES found on upper
leaflets only

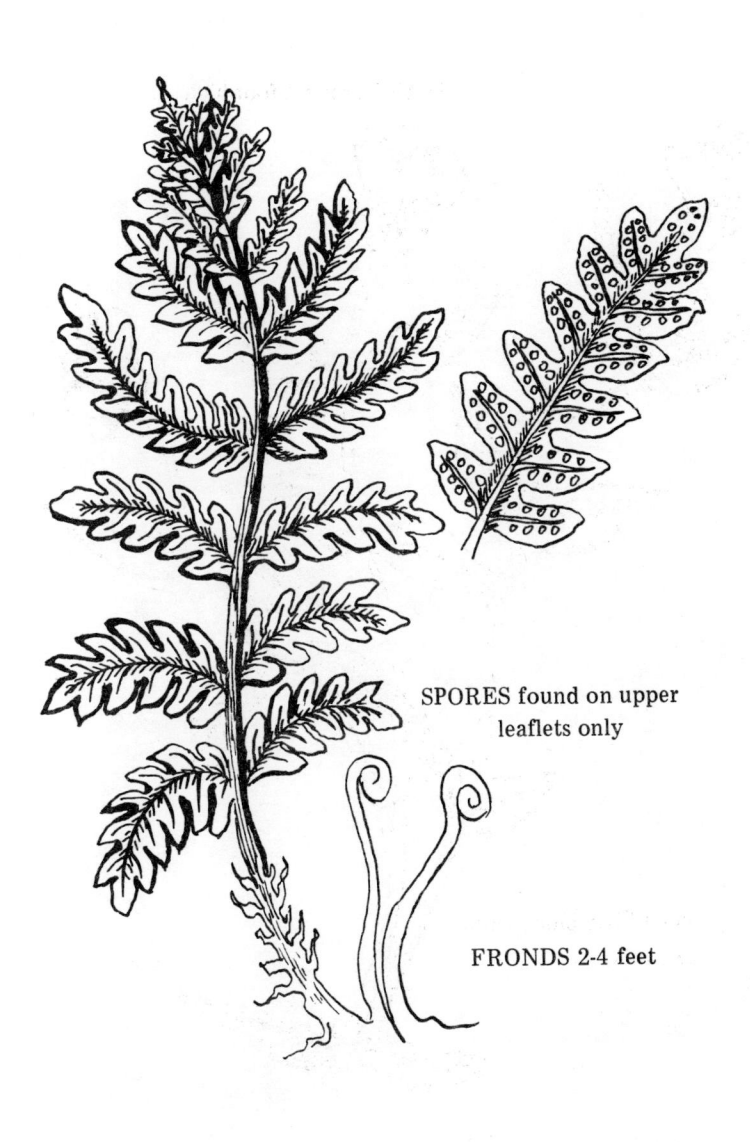

FRONDS 2-4 feet

CRESTED WOODFERN

PLANT grows 1 foot high

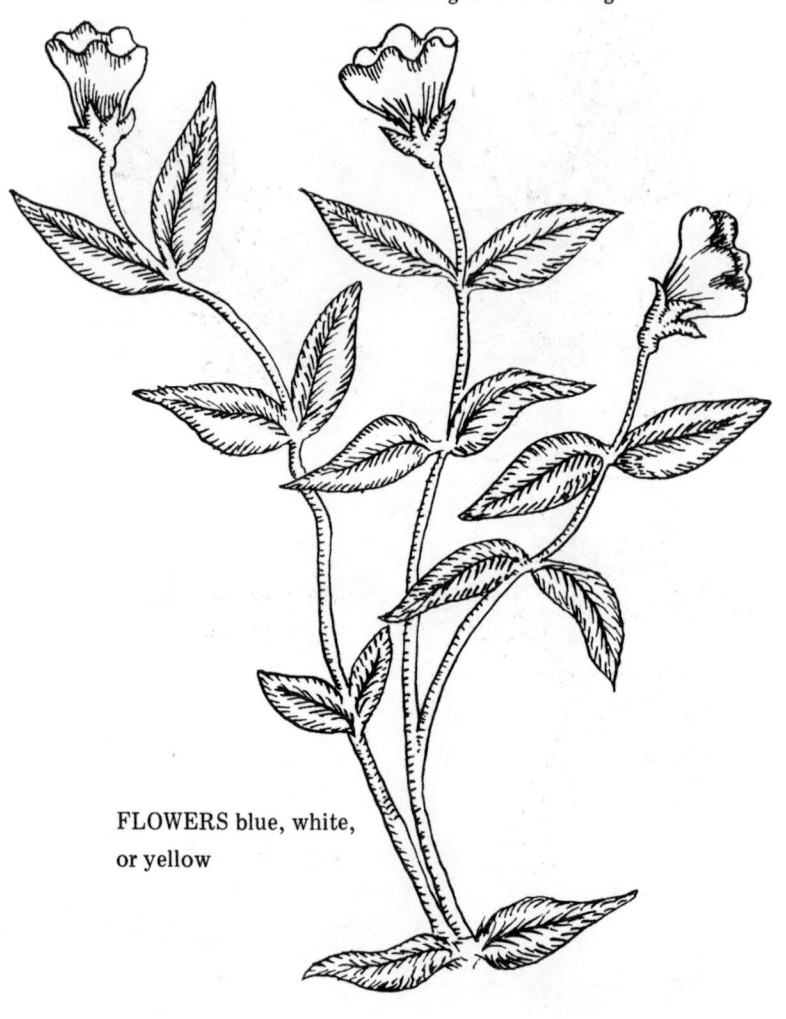

FLOWERS blue, white,
or yellow

HEDGE HYSSOP

HEIGHT 6-18 inches

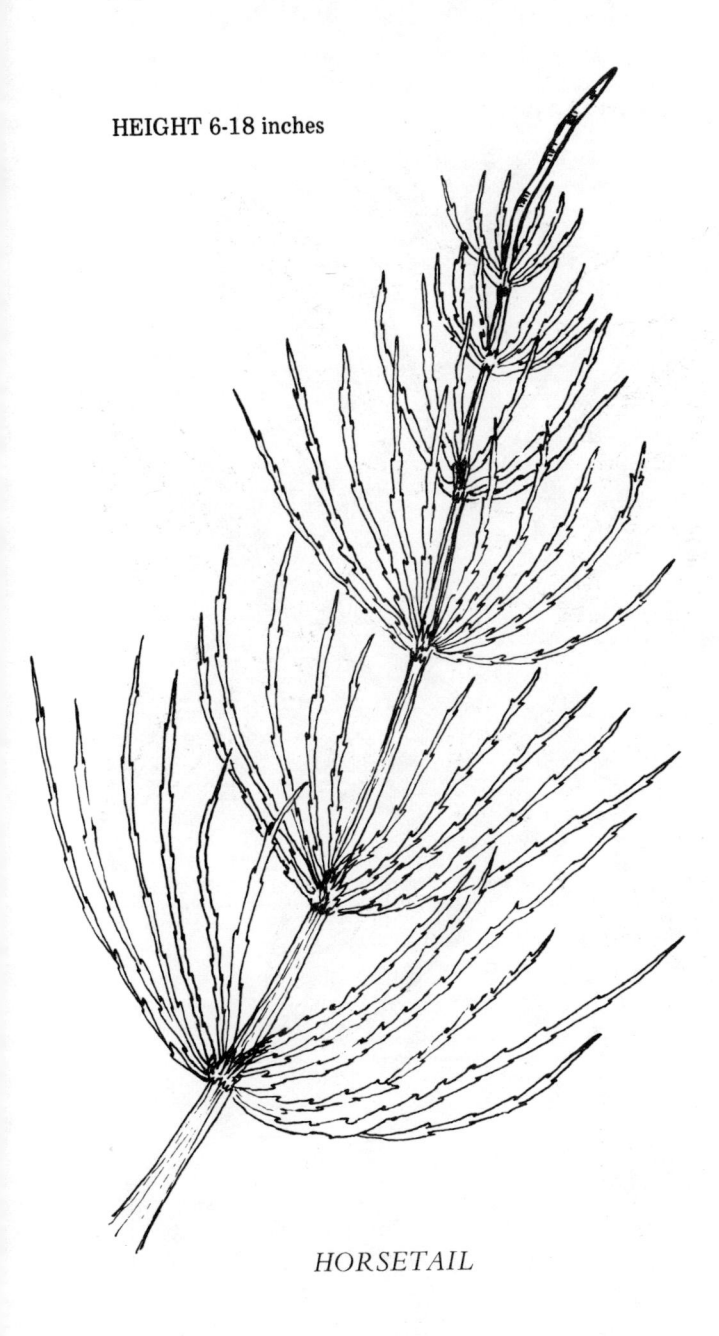

HORSETAIL

BUSH grows 2-4 feet

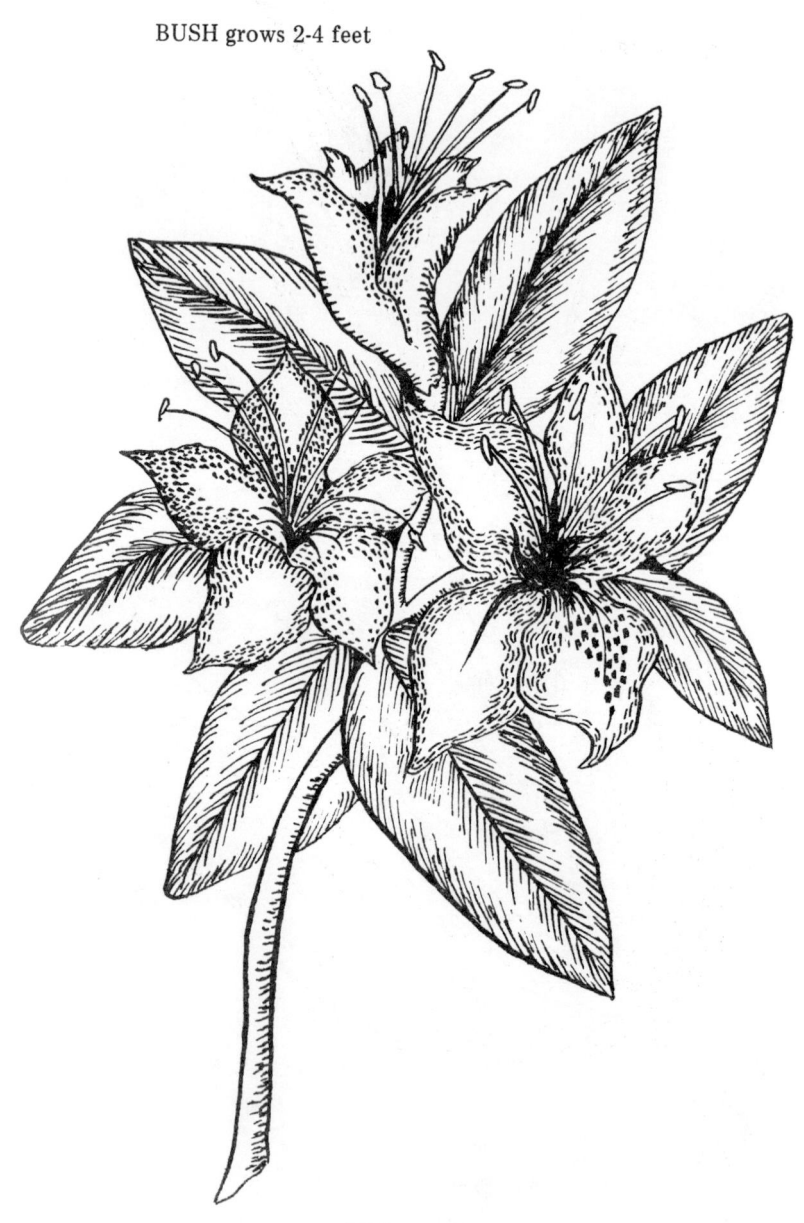

WILD AZALEA

20-50 feet

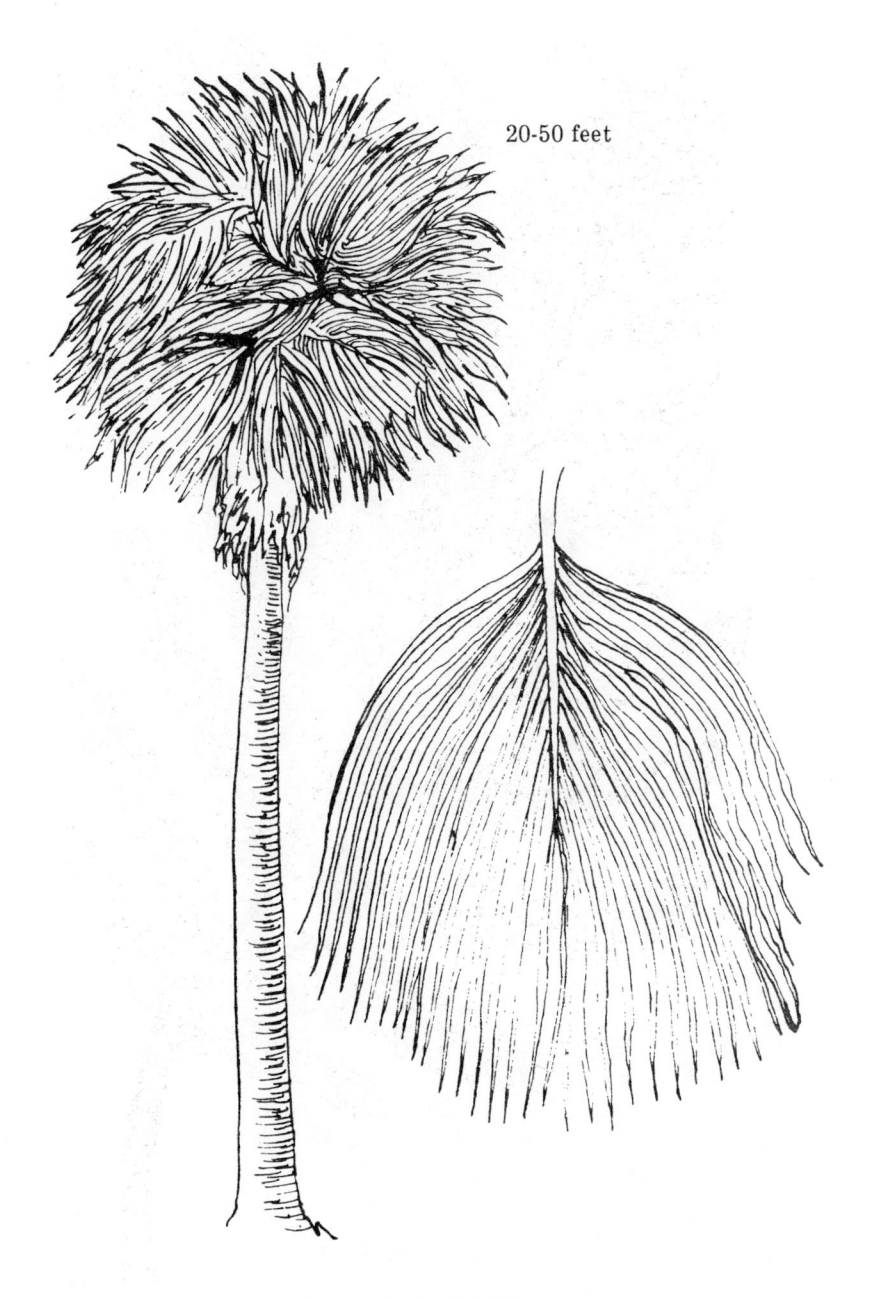

CABBAGE PALMETTO

SPORES

Fertile Leaf edges curled

veins

Nonfertile Leaves

MARSH FERN

LEAVES 3-6 inches in diameter

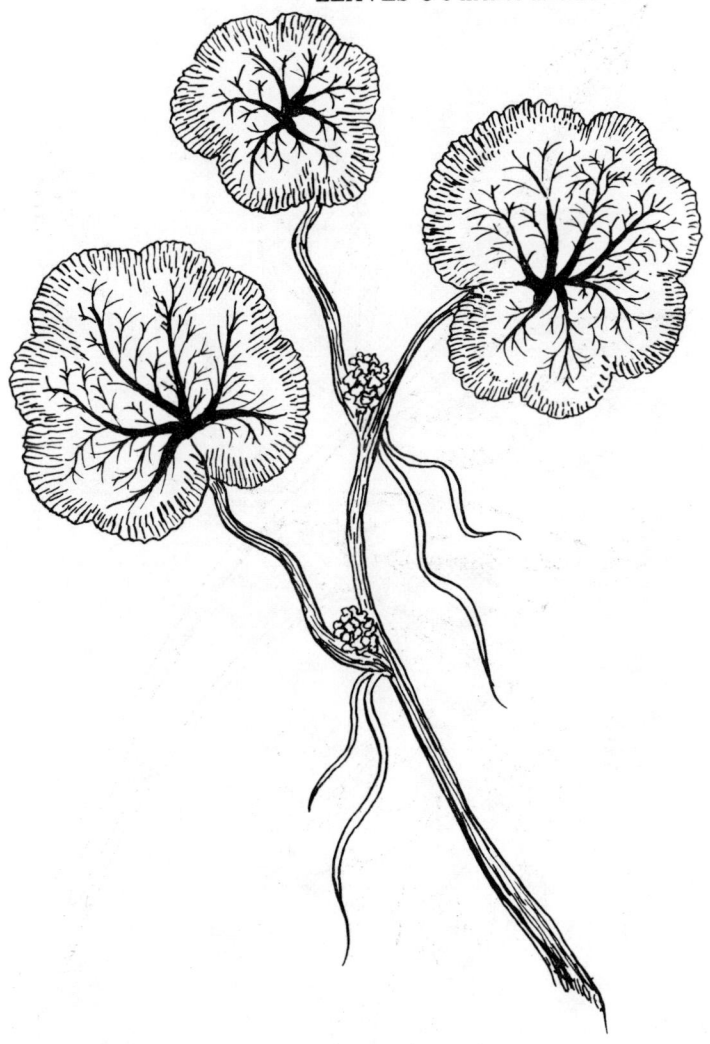

WATER PENNYWORT

GROWS to 20 feet

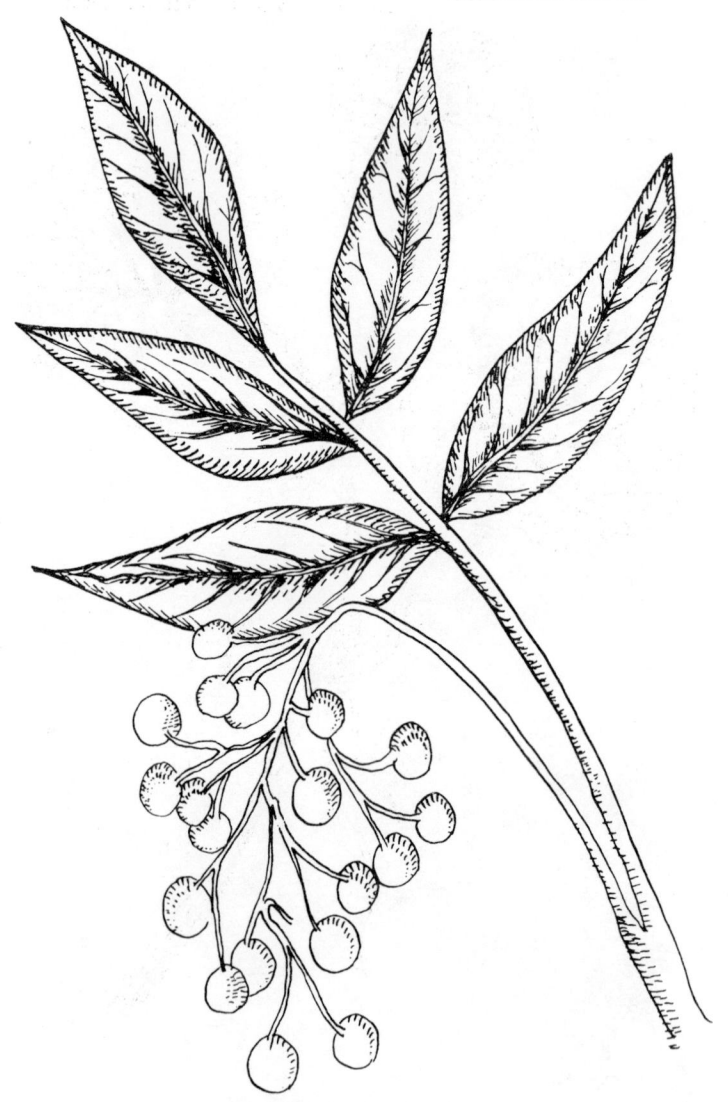

POISON SUMAC

SPORES

RATTLESNAKE FERN

SPORES

ROYAL FERN

LEAVES 1-3 feet long, 6-12 inches wide

SKUNK CABBAGE

STEMS 3-6 feet long
LEAVES 6-12 inches

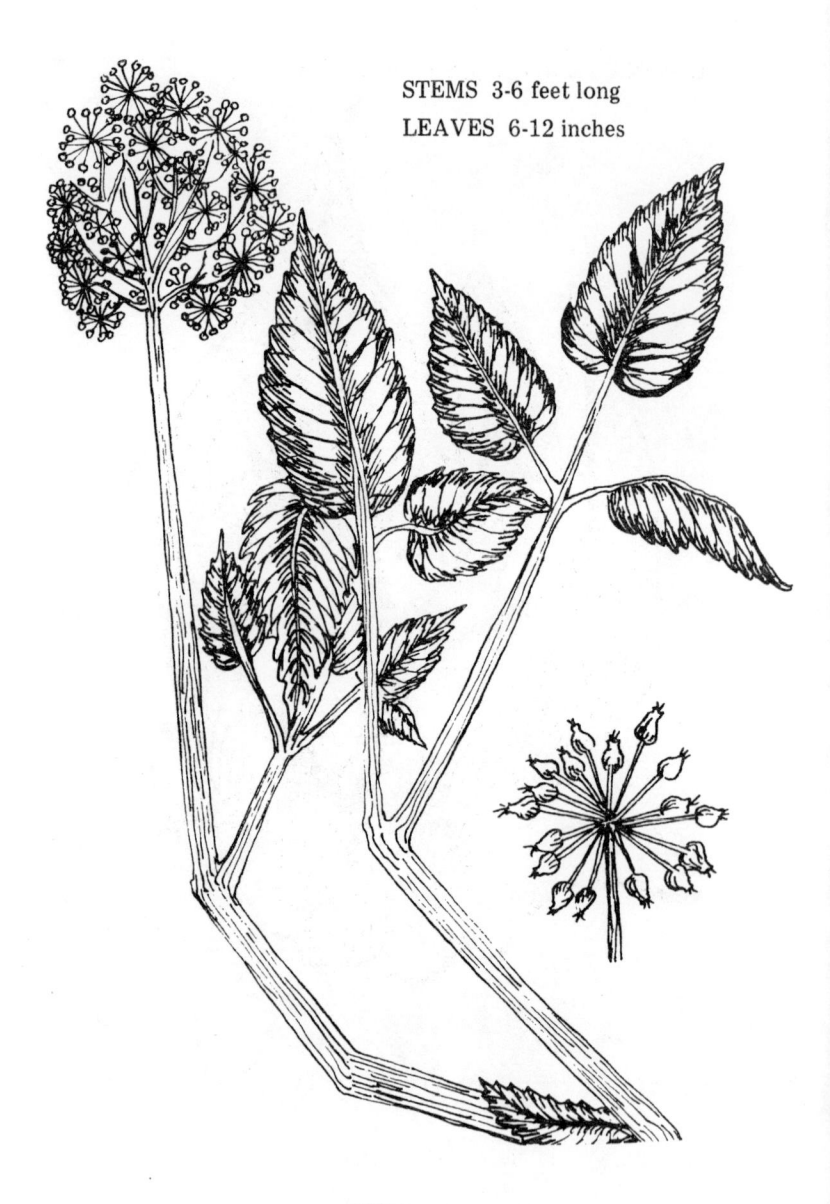

SPIGNATE

20-60 feet

WILLOW

FRONDS 2-4 feet

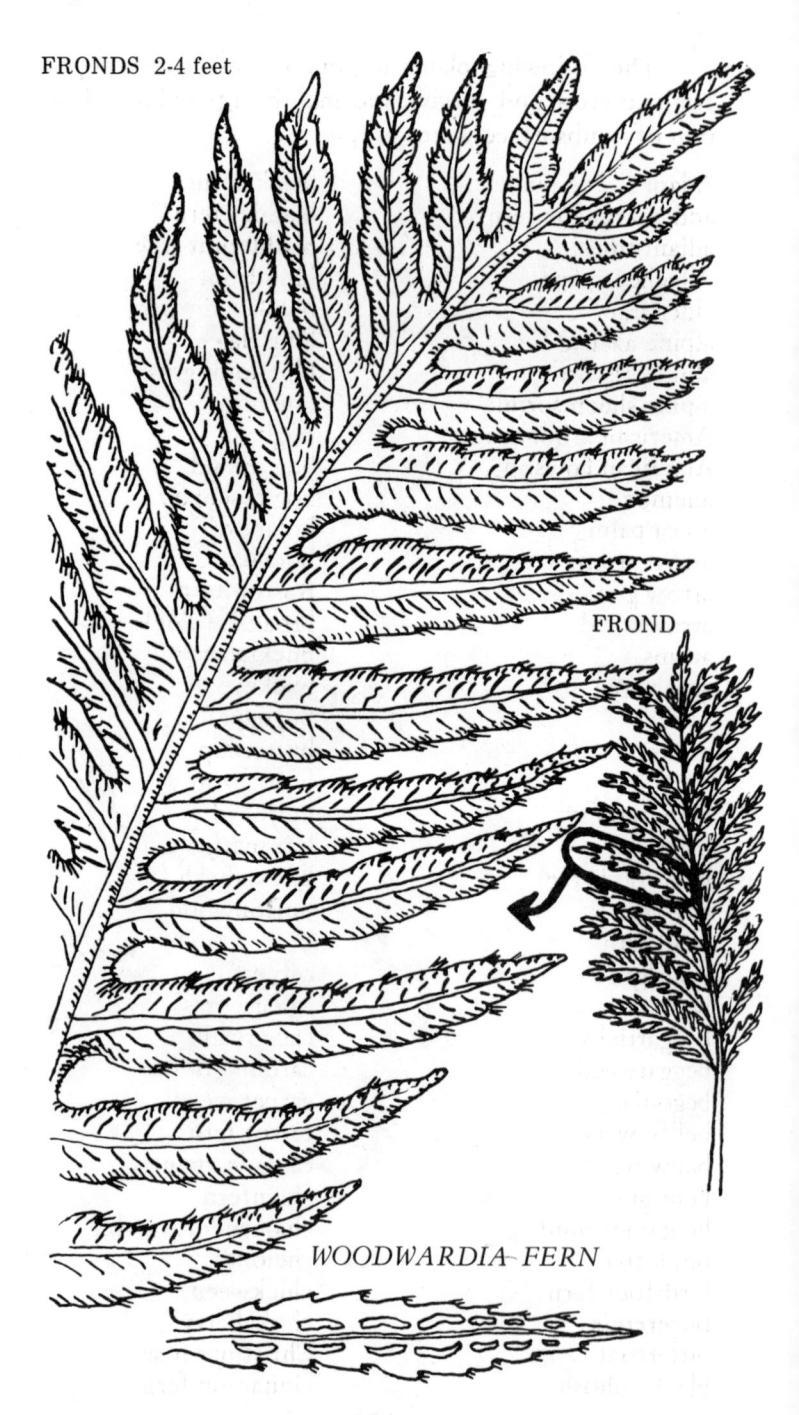

FROND

WOODWARDIA FERN

The following plants require a moist, often boggy earth to grow and survive and may be a possible indication of a subsurface water supply:

Adam & Eve plant
adder's tongue fern
adiantum
Allegheny vine
alder tree
alpine azalea
alpine bistort
alpine sheep sorrel
American arborvitae
American cowslip
anemone
areca palm
arrow arum
arrow grass
arrowwood
arums
aspen tree
aster
avens

bald cypress
bamboo
basswood tree
bay tree
bead fern
beardgrass
bedstraw
beggarticks
beggarweed
begonias
bellflowers
bellwort
bent grass
bergamot mint
birch tree
birdsfoot fern
bittercress
bitterroot
black cohosh

black spruce
bladder fern
bladdernut tree
bloodroot
blue-eyed grass
blue flag
blue joint grass
bluet
bogbean
bog orchid
bog rosemary
bolandra
borage
Boston fern
broadleaf dock
buckbean
buckeye
bugbane
buttercup
butterwort
bur head
bur marigold
buttonbush
buttonwood

calypso
canary grass
cane reed
cardinal flower
carpet weed
carrot fern
cascade violet
chainfern
chamaedorea palm
chelone
chickweed
chokeberry
Christmas rose
cinnamon fern

citronella
climbing hemweed
clintonia
club rush
clustered dock
coffee-fern
colic root
columbine
common fern
common giant fennel
cotton grass
cow parsnip
cowslip
creeping charlie
creeping jennie
creeping snowberry
cress
crowfoot
cuckoopint
cucumber root
cudweed
curly dock
curly grass
currant
cushion fern

daisy leaf violet
dayflower
decumaria
deer cabbage
deer fern
dogwood
douglasia
dragonhead

earth almond
elderberry
elder tree
elephant's head
elk clover
elodea
enchanter's nightshade
everlasting pea

false alum root
false hellebore
false spirea
fan palm
field mint
filmy fern
five-finger fern
flowering fern
flowering raspberry
flowering rush
forget-me-nots
fothergilla
fragrant fern
fringed orchid

gentian
globe flower
golden club
golden-eyed grass
golden larch
goldenpert
golden polypody
golden saxifrage
goldenseal
Goldie's fern
goldthread
gooseberry
gormanias
grape fern
green cliff brake

hairgrass
hardhack
Hartford fern
hartstongue fern
hedge nettle
hellebore
hemp
hibiscus
hobblebush
hooded lady's tresses
hornbeam tree
hot-springs panicum

36

huckleberry

Indian cucumber root
Indian paintbrush
Indian pipe
insectivorous plants
interrupted fern
ironweed

jack-in-the-pulpit
jewelweed
johnny-jump-up
juncus

Kenilworth ivy
knotgrass

Labrador tea
lace fern
lady fern
larch tree
large-leaf lupine
leather leaf
licorice bush
linaria
linden tree
lizard's tail
lobelia
Longbeach fern
lovage
lovegrass
ludwigia

mahonia
maidenhair fern
manna grass
marsh cinquefoil
marsh clover
marsh fern
marsh mallow
marsh marigold
marsh violet
Massachusetts fern

masterwort
may apple
meadowfoam
meadow foxtail
meadow knotweed
meadow rue
meadowsweet
Mexican tea
moneywort
monkeyflower
monkshood
moonseed
mountain avens
mountain heath
mountain holly
mountain maple
mousetail
musk plant
musquash root

navelwort
nettle
nightshade
ninebark
northern bog violet
Nuttall's quillgrass
nut grass

oak fern
Oregon grape
osier tree
osoberry

paintbrush
papyrus
parnassia
parnassus
paw paw
pearlwort
pellonia
pennycress
pennyroyal
peppermint

pillwort
pitcher plant
plantain
poison sumac
possum haw
primrose willow
purple cliffbrake
purple loosestrife
puttyroot

queen lady's-slipper

ranunculus
red birch
red robin
redtop
reedgrass
rhodora
rock jasmine
rose gentian
rose mallow
rue anemone
rushes
rustyleaf

salal
salmon berry
satin poppy
saxifrage
saw fern
scratch grass
sedges
semaphore grass
senecio
sensitive fern
shield fern
shooting star
shrub yellowroot
silverweed
silvery spleenwort
skullcap
smartweed

snailseed plant
snake mouth
snake root
snake weed
Solomon's seal
sour gum tree
spearmint
speedwell
spicebush
spiderwort
spike rush
stargrass
starwort
steeple bush
St. Johnswort
Stokes' aster
stoneroot
sticktight
strap fern
studflower
summer-sweet
sun cup
sundew
swamp dewberry
swamp laurel
swamp locust
swamp loosestrife
swamp maple
swamp onion
swamp pink
swamp rose mallow
swamp whitehead
swamp white oak
sweetbells
sweet flag
sweet gale
sweet pepperbush
sweet pitcher plant
sweet spire
sycamore tree

taper fern

38

tea fern
thalia
tree square
tickseed
toad lily
toad rush
toothcup
touch-me-not
traveling fern
trillium
trumpets
tuber fern
tulip tree
turtlehead
tupelo
twayblade
twinflower
twinleaf
twisted stalk

umbrella leaf
umbrella palm
umbrella pine
uvularia

veronica
vine maple
viola

walking fern
wall fern
walnut fern
wandering jew
watercress
water elm

water hemlock
water hoarhound
water parsnip
water plantain
water purslane
waterweed
wax myrtle
western spirea
white birch
white cedar
white violet
wild blue flag
wild cranberry
wild fuchsia
wild ginger
wild ginseng
wild horseradish
wild quince
wild rosemary
wild rye
wild sweetpea
wild wisteria
willow dock
willow herb
willow oak
wind poppy
wintergreen
witch grass
withe rod
woodruff
woodrush
woodwardia fern

yellow cress

Once you have found the water source you plan to use, it is a good idea to have this water analyzed before the property is purchased. This may not be necessary when you are buying virgin land in sparsely populated areas. It may seem an unnecessary hassle to test the water, but you will save much trouble, time and expense in the long run, if the water turns out to be impure. Perhaps the seller will agree to undertake the task and expense of having the water tested. Without good water, the property would be of little value for a homestead.

If the water contains poisonous pesticides, it is unfit for any use. If it contains high amounts of iron, salt, manganese, boron or nitrates, it should not be used for drinking or cooking.

If the water has a reddish tinge, it could be an indication of a high iron content, or the presence of tannic acid. Both are undesirable. The iron would come from a high concentration of this mineral in the soil. This land should be avoided. If it is tannic acid, it may have come from vegetation containing it, such as tan oak or sassafras trees. Rainwater may have leached them and carried the acid to the water system. The presence of tannic acid is possible to remedy by the addition of limestone to the water supply to neutralize it, but it is still more desirable to begin with pure water.

Impure water generally has a distinct odor and taste. If you have a suspicion that the water is impure, have it tested.

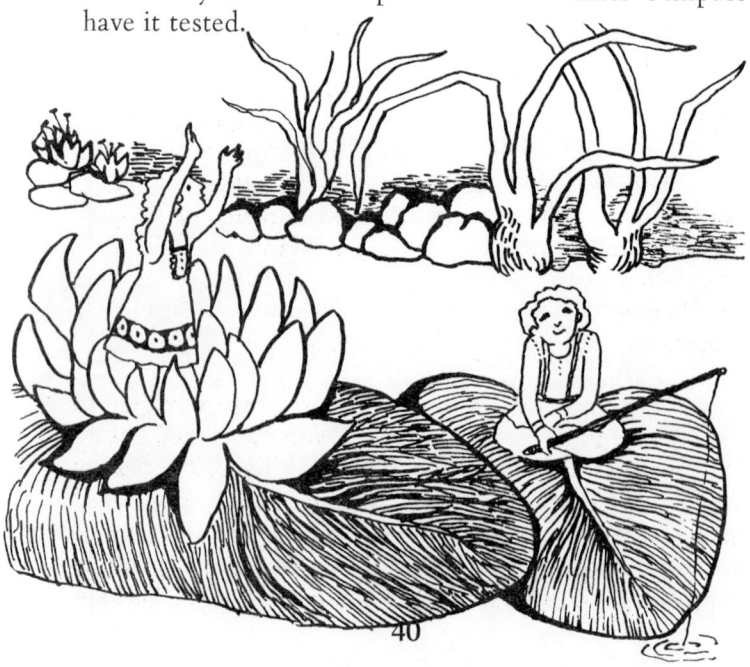

DIRECTION OF SLOPE

The direction of slope is determined by the direction you are facing when you look downward from the top of a hill. On the north slope, the land slopes downward toward the north, and on the south slope the land slopes downward toward the south.

In hilly or mountainous areas, land should be selected that has large benches of south or east sloping land, or good valleys with sunny exposure that have at least two feet of good topsoil. Without these conditions, it is very difficult to develop a suitable farmstead on mountainous land. It is advantageous to plan for your garden to be on a gentle slope, as sloping land allows for proper drainage and is beneficial in preventing frost damage.

If the land you are selecting is in a cold area where there are early fall and late spring frosts, it is best to choose land on a southern slope of a hill. Here there is a higher intensity of heat and light from the winter sun, which provides a warmer earth in the winter, lengthening growing season considerably. On level areas, and especially on a northern slope, an equal amount of sun ray energy is spread over a wider ground area, causing diffusion of light and heat. Ground will remain colder throughout the winter and will take longer to warm up in the spring.

An eastern slope is also suitable for a farmstead, although the intensity of warmth will not be as great as on a southern slope. Length of hot and cold seasons may be a deciding factor. Areas with long, hot summers could do without the intense afternoon heat of the southern slope, whereas areas with long cold winters would be greatly benefited. An eastern slope catches the earliest morning sunlight, which warms up houses and greenhouses quickly. Also, the garden will get the

DETERMINING DIRECTION OF SLOPE

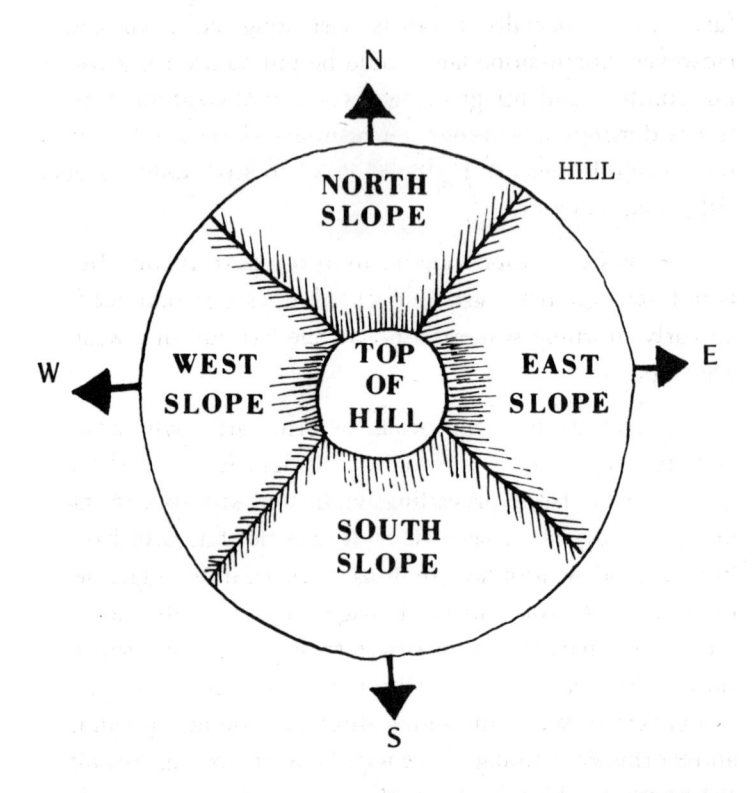

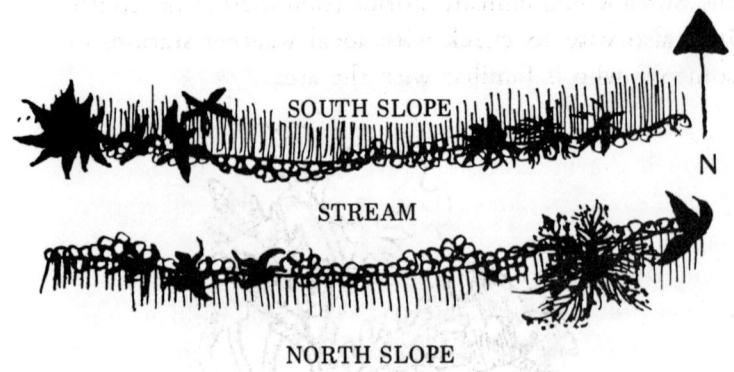

- NORTH SLOPE
- WEST SLOPE
- TOP OF HILL
- EAST SLOPE
- SOUTH SLOPE
- HILL

SOUTH SLOPE

STREAM

NORTH SLOPE

early sun when it needs it the most to warm the earth after a cold night.

A northern slope would not be very good for a farmstead, especially in areas with long, cold winters. However, north-slope land could be put to use for growing conifers and for growing certain cool-weather vegetables during the summer. A northern slope would also be a suitable place for a pond stocked with cold-water fish, such as trout.

A western slope can be used for pastureland, but is not very good for growing a garden, as a garden needs an early morning sun, rather than the late sun of a western slope.

If you are near the ocean or other areas where the climate is temperate, the main consideration as to slope is protection from prevailing winds and storms coming in off the ocean. It is best to locate the farmstead behind a good windbreak, such as forest, hill, or ridge, because these storms can be damaging to orchards and to crops. Storms usually originate from one general direction in an area. By observing the tops of tall trees you can generally see from which direction the heavy winds and storms are coming. Tree tops bent or curving toward the north would indicate storms coming from the south. It is also wise to check with local weather stations or someone who is familiar with the area.

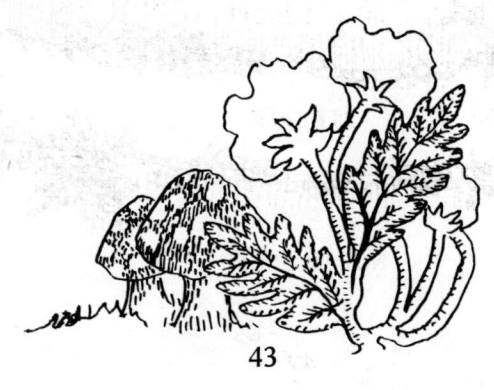

LIGHT FROM THE WINTER SUN FALLS MORE DIRECTLY, THUS MORE INTENSELY, ON THE SOUTHERN SLOPE

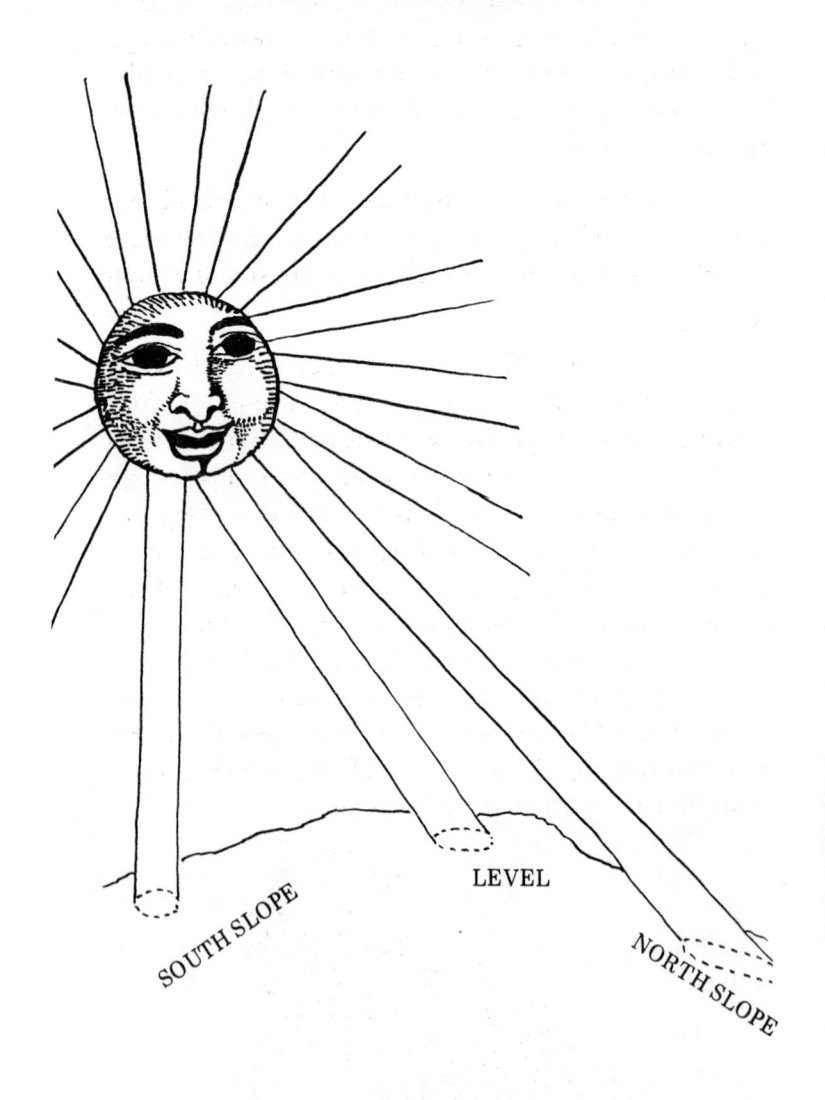

TYPE OF SOIL

In soil we find the Roots of Life. Working harmoniously with the earth will improve it, just as nature carries on continual renewal by returning vegetation, rocks and water to the earth at the end of each life cycle. Much may be learned about Nature's ways through observation of her elements.

Whatever grows reflects the quality of the earth. Whatever you cultivate will reflect the quality of energy you have put into the materials you have to work with. Conservation of energy has always been important and we have seen what happens when energy is misdirected and wasted. This applies on many levels and, back on the farmstead, it applies in connection with the soil. By beginning with the best soil possible, you will save much energy and produce better results. The energy that you would have put into the soil to make it merely workable will go toward making it abundantly fertile.

There is much to know about working with Nature. By understanding her language, her signs and symbols, you will discover new dimensions of working with and within the earth.

BASIC ROCK FORMATIONS

Igneous: Rock substance was at one time molten and forced upward from deep down in the earth. This rock formation forms a good basic soil for growing a garden. It includes: basalt, (granite makes poor soil).

Sedimentary: Rock substance was formed by various minerals carried down and deposited in layers by the elements, such as wind, water, glaciers, and the decomposition of plants. This type of rock often contains fossils. This is a good basic rock which forms soil rich in minerals for growing. It includes: sandstone, shale, limestone, peat, and dolomite.

Metamorphic: Rock which has undergone change within the earth through heat and pressure. Marble, a metamorphic rock, was once limestone. Slate, also a metamorphic rock, was once clay. This rock formation usually forms a poor basic soil for growing. It also includes: quartzite, schist, mica, gneiss, and hornfels.

SOIL COMPOSITIONS

Alluvial: Soil which has been carried and deposited by water, winds, and other natural elements. It is generally found in lowlands near creeks and rivers. This soil is usually composed of varying amounts of clay, sand, and silt. It generally makes an excellent garden soil, being rich in minerals and having good drainage.

Clay: Texture tight and nonporous. It is hard to work, as it forms hard lumps when dry which crumble to a fine powder when crushed. Sticky when wet and holds water on surface. Red to tan coloring, depending on amount of iron; sometimes bluish or gray. Clay color varies according to mineral content. Solid clay is a poor soil for growing a garden because its tight texture does not allow for proper drainage and aeration. The addition of one part sand and one part silt to one part clay will transform it into good garden soil. Clay soil is excellent for building a pond, as it is extremely nonporous and will hold water well. Very heavy clay, usually found in arid regions, is called adobe. This is difficult, but not impossible to work. It makes good building bricks. Soil containing clay may cause many problems with sewage system drainfields because of the poor aeration and drainage qualities of clay.

Forest Soil—Coniferous: Pine, fir, western cedar, and other cone-bearing trees. Generally a poor, high-acid soil. Will need compost and lime to improve quality and texture for growing.

Forest Soil—Hardwood: Oak, maple, walnut, bay, madrone, and other hardwood trees. Generally soil is better than coniferous forest soil, but may still be rather acid (especially oak) and will need compost and lime to balance soil for growing. Once acid has been leached from fallen leaves by winter rainwater, the leaves will make an excellent mulch.

Grassland: Generally a good soil, unless clay content is very high. Much better for growing than forest soil. Quality, uniformity, and thickness of coverage of the grasses growing on the land tell about soil quality. Good soil produces high-quality grasses covering thickly and uniformly.

Humus: Dark, extremely rich soil material formed by the slow decomposition of organic matter. This is excellent soil for a garden.

Loam: A rich soil composed chiefly of silt, sand, and clay with some decayed vegetable matter. This is very good soil for gardens, as it contains many valuable nutrients and allows for good drainage, being not too porous yet not too tight.

Sandy Soil: Ranges from extremely porous to medium porous, containing relatively large particles of rock. Sand is heavy in weight as compared with silt. Works easily, has good aeration. Soil with high sand content allows minerals and other nutrients as well as water to drain out quickly, so for a garden it is wise to add plenty of humus and silt. Sand is very beneficial for increasing the porosity of tightly compacted soils, such as clay.

Residual Soil: Soil formed by the gradual breakdown and decay of rock, slowly building up on the surface to create topsoil. Unlike alluvial soil, it remains on surface of parent rock. Fertility depends on type of rock it is formed from, plant humus content, climate, and depth of soil. If it is too deep, nutrients leach out quickly. This soil has good drainage.

Silt: Composed of minute particles of rock, similar to fine sand, and humus, carried and deposited by slow-moving water. It is fairly porous and is much lighter in weight than sand. Makes up a large part of alluvial soil. River and creek silt is generally very rich and has a good texture.

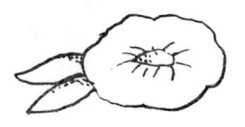

Finding fertile soil is simple once you know what to look for. Vegetation will tell much about the quality of the soil. Lushness of plants, grasses and trees is indication of rich soil. The more abundant, healthy and dark green the foliage, the more fertile the soil. In meadows, evenness of grass covering is also a good sign. Scrubby, uneven cover of plants and grasses is a sign of poor soil. Specific types of plants grow on various soils, as different plants have their own special nutritional needs. The following lists will give you a general idea of what to look for. These should be taken into consideration along with other local conditions. For instance, plants with thorns, and thistles, grow under conditions that make growth difficult, such as poor soil or lack of water, and so must produce thorns to discourage various vegetarian browsers that might prevent propagation of the species. Finding thistles growing in an area may mean poor soil or lack of water, if there are other conditions that agree. On the other hand, if there are thistles growing along with plant and other indicators of rich soil, the thistle theory may not hold true.

INDICATORS OF FERTILE SOIL

acacia	magnolia
arum	mandrake
bent grass	moonseed
Bermuda grass	nettles
begonia	obedient plant
bishop's cap	pampas grass
bloodroot	paw paw
blue cohosh	pigeon cherry
blue-eyed grass	pink root
canyon poppy	quake grass
celandine	sea holly
chicory	shinleaf
citronella	shore grape
clintonia	southern fox grape
clover	starflower
coneflower	stoneroot
cress	striped maple
Culver's root	timothy
desert candle	trumpet vine
dutchman's breeches	toothed woodfern
elderberry	tulip tree
flowering rush	twin-leaf
giant ragweed	twisted stalk
goldenseal	uvularia
gooseberry	walking fern
holly	walnut tree
hollyhock	waterleaf
horse balm	whitethorn
horseweed	wild aster
jack-in-the-pulpit	wild ginger
Jacob's ladder	wild lilac
Kentucky coffee tree	wild oats
larkspur	wild onions
legumes	yellowwood
liverleaf	

Grasses growing thickly and covering ground evenly. Trees growing straight with smooth, long, even trunks.

INDICATORS OF NONFERTILE SOIL

amaranthus – the more brilliant the color, the less fertile the soil.

American smoke tree

aspidistra — one variety is striped yellow and green in poor soil, turns green in rich soil, losing yellow stripes.

bayberry	henbane
bishop's weed	maclura tree
blue weed	poppy mallow
brake fern	redwood tree
broom crowberry	Scotch broom
chittam wood	thistle
cosmos	thoroughwax
fir	vipers bugloss (flowering)
gum plant	wild hyacinth

Grasses growing thinly and in uneven patches.
Over-sized roots on undersized plants.
Trees growing with misshapen, spindly trunks.

Vegetation growing locally also tells much about the acidity or alkalinity of the soil:

HIGH ACID

berry vines	manzanita
butterfly weed	moss
dutchman's breeches	mountain laurel
evergreens	orchid
ferns	pecan tree
flax	pine tree
heath	redwood tree
heather	sorrel
hickory	spicebush
lady's-slipper	spruce tree
lily	wild azalea
lupine	wintergreen
madrone	yew tree
magnolia	

LOW ACID

apple tree
gooseberry
larkspur
mustard
oats

queen lady's-slipper
rye
vetch
wild grape

NEUTRAL—ALKALINE

alfalfa
clover
iris

quince tree
wild blue flag

TESTING THE SOIL

Just as it is wise to test the water before purchasing land, it is wise to test the soil. This is mainly important if you are buying land that has been used before or land near industrial centers. Virgin land needn't be tested unless you suspect the presence of harmful minerals.

Some soils contain minerals that make the soil unsuitable for growing food crops, and some contain minerals that will provide an extremely rich soil. It's good to know what you have to work with in the beginning.

If soil contains pesticides or high amounts of salt, arsenic, boron, iron, mercury, or lead, it is unfit for growing.

It is very good to find lime or magnesium in the soil. They are beneficial for growing a garden, as they tend to make the soil more alkaline.

To test the soil, take it to an independent laboratory. They have the necessary equipment to test for these minerals, equipment not generally found in a do-it-yourself test kit.

TESTING THE SOIL COMPOSITION

Appearance and texture tell much about the composition of the soil. Coloring is important in determining richness as well as mineral content.

Dark Brown, Loosely Packed Soil is just about the best possible soil for growing. Its darkness is attributed to its high humus content. Drainage and mineral content are excellent.

Tan, Loosely Packed Soil which is alluvial is good, as it is high in mineral content as well as small particles of organic matter. It will have good drainage if it contains much sand, but will need the addition of compost to meet nutritional requirements.

Tan or Red, Tightly Packed Soil is probably high in clay content. Contains plenty of minerals but will need sand and/or organic matter such as compost to loosen porosity to provide aeration and drainage.

Dark Red, Tightly Packed Soil is probably heavy clay with a high iron content. This type of soil should be avoided, as high quantities of iron in soil cause impurity of water.

Blue, Tightly Packed Soil is generally heavy clay. Terrain containing much of this has a tendency to erode and change by sliding, especially in heavy rain areas. With the use of compost and materials such as sand, silt, and leaves, may be developed into a fair garden, provided that the garden doesn't slide downhill.

Gray, Tightly Packed Soil found in hot, arid regions contains very little humus and may be extremely alkaline. Composting and leaf mulch will improve soil somewhat.

Feel the texture of the soil. If:

Crumbly when rubbed between fingers when moist it contains silt or sand.

Grainy in texture it contains silt or sand.

Settles Quickly when stirred into a glass of water, it contains sand and possibly silt.

Settles Slowly when stirred into a glass of water, it contains clay.

Sticky when wet, it contains clay.

"Ribbons" like toothpaste out of a tube when wet and squeezed between fingers, it contains clay.

Hard Lumps form when dry, it contains clay. The more difficult it is to break up the lumps, the higher the clay content.

Smooth and Greasy to the touch when wet, indicates high clay content.

Compacts when wet or dry, it contains clay.

TESTING THE SOIL DRAINAGE

This test is very useful for testing the water-holding capacity of earth for planting a garden, digging a pond, putting in a foundation, and many other projects. It can also give an indication of the basic type of soil, such as clay or sand.

1. Take a gallon can with holes punched on the side near the bottom.
2. Place a 1-inch layer of small gravel mixed with coarse sand in the bottom.

3. Fill can 2/3 full with wet soil to be tested, and lightly pack soil into can with fingers.

4. Pour 1 quart water into can. (Place a piece of cardboard on top of soil to prevent water from eroding a hole into the soil surface.)

If water drains within 15 minutes soil is very porous.

If water drains in 15-60 minutes soil is moderately porous.

If water does not drain within 24 hours soil is nonporous.

It is best to repeat step 4 of the test once or twice to be certain of the results.

ROADS

Land near a county road has both advantages and disadvantages. Location near the road makes the land more accessible, especially during winter months. It provides better possibilities for roadside vegetable and livestock marketing. Not having to travel through another person's property to go to and from your home has its obvious advantages. Friends will probably drop by more often, but then so will just about anyone who happens to be out of water, gas, or a place to stay. There may also be the problem of "borrowers" who take lumber, tools, etc., and forget to return them because you "weren't using it anyway". There are obvious pros and cons on both sides and you must decide for yourself what you are looking for. If it is privacy, the further back the better.

Access roads are very important for the obvious reason of getting you, your building and other supplies, groceries, etc., to your land. Many people have bought land, assuming they had use of a road, and then found themselves "land-locked" with no legal access, a locked gate, and uptight neighbors. Be certain that all rights of way are clearly marked, on the land itself as well as in the deed.

If there is no road to the land, it is important to consider the feasibility, in terms of terrain and expense, of putting one in. If possible, try to get the person you are buying from, such as the realtor, to have the road cut. It is best to supervise this yourself so that you get the road exactly where you want it. If you are going to cut the road yourself, be sure you know what you are doing. Either way, read Chapter II, section on Roads.

CLIMATE

In Climate we find the Breath of Life. During the early development of the earth, animal and vegetable life grew and evolved according to the climatic conditions. Even rocks and the basic earth formations were affected by climate. We ourselves are affected by the climate in subtle ways that we aren't always conscious of. When selecting land, choose an area with a climate you can live with. Temperate, 70-degree weather year 'round is not for everyone, nor are the seasonal areas of extreme heat and extreme cold. It will be possible to alter and improve climate through development of microclimates, and to lengthen the growing season. Then, too, there are small microclimatic regions where, within the space of a hundred miles, you will find all four seasons year 'round. It is important to be aware of the advantages as well as the disadvantages of the various types of climate. Through careful observance of vegetation in the area, how it grows, what changes it goes through, coloring and thickness of growth, much may be told about the climate in the area.

Temperate: Temperate zones are areas where there is little change in the climate from season to season. These areas are usually found on the coast near the ocean, where the ocean controls a microclimate on a large scale. Just a few miles inland the climate may change drastically, especially when there is a high mountain ridge between these inland areas and the ocean.

Summers in a temperate zone are generally cool, winters mild, which creates a longer, sometimes year 'round growing season. These areas are more vulnerable to storms and hurricanes, so buildings should be built sturdily. Humidity here will also be high. Generally, water is plentiful and soil will be of good texture.

Hot Year 'Round: These areas will be good for year 'round growing of vegetables, fruits, and citrus fruits, but be certain that there is plenty of water unless you just want to raise cacti. Houses will need insulation against heat. Insects and rodents may be a problem. Ponds or other swimming areas may be a necessity.

Cold Year 'Round: Poor for growing outdoors, but greenhouses may be used to great advantage as long as there is sufficient sunlight. Fuel costs (in terms of time, energy and money) may be high. House will need to be built tightly with good insulation. Consider tools such as chainsaws, snow equipment, livestock shelters.

Hot and Humid Year 'Round: Areas that are continuously hot and humid all year 'round encourage rapid plant growth and lushness, but the disadvantages and discomfort of living in these areas are great. Hot, humid climates breed vast quantities of detrimental insects, such as flies, mosquitoes, gnats, roaches, fire ants, ticks, poisonous spiders, and many garden-eating bugs. Poisonous snakes are also abundant in the hot, humid areas, but they are not as potent as the ones found in the hot, dry areas. Concern for heating may not be a necessity, but air conditioning may. General health of people can be affected by living in these regions. The idea of an enchanted tropical island appeals to many people, and for some it may be the perfect environment. There are often wild foods growing so abundantly that there is no need for cultivation of a garden.

Hot Summers: Good for growing crops as long as water is plentiful. Ponds, water reservoirs and swimming areas will be a consideration. Insulation of house to keep out heat will be important. Refrigeration to prevent food spoilage may be a consideration.

Cold Winters: Some crops may still be grown, both outside and in a greenhouse. Fuel is an important consideration, as well as house insulation to avoid heat waste. Getting supplies in should be considered. Animal dwellings in harsh weather may be necessary.

Hot Nights: Good for growing plants, as plants do much of their growing at night. Insects love hot nights, too, however. Hot nights can make sleeping uncomfortable.

Cold Nights: Not too good for growing plants; as mentioned above, they like warm nights for growing. Possible to set up a microclimate to store daytime heat for night. Fires for warmth may be necessary, even in summertime, at night and early morning.

Heavy-Rain Seasons: Nutrients may be leached from soil, so much compost and manure should be added to garden after rainy season ends. These areas generally have a high natural water supply, which is good. Best to fill ponds and other reservoirs during this season for use during drier seasons. House must be built leak proof and have good drainage around buildings and basements. Roads may need careful work and constant attention to prevent washing out. Good supply of firewood will be needed to keep place warm and dry. Animal shelters may be needed. Building must be done before rainy season begins. Four-wheel-drive vehicle may be necessary to travel up and down the road. Waterproof clothing will be a necessity.

Very Dry Seasons: Water will need to be stored during the rainy periods for use during these dry seasons. Even so, it may be important to conserve water carefully during the dry time. This is not the most favorable condition under which to grow a garden, though it can be done with careful storage and conservation of water. It will be good to develop bodies of water and other natural microclimates to help raise humidity of the area. Always try to keep a store of several gallon jars or 55-gallon drums filled with water in case of an emergency. This climate will be good for drying fruit.

Thunderstorms: Some areas have violent thunderstorms and hail. Both can be very damaging to crops, especially hail. It is best to check with nearby weather stations to find out the frequency and severity of these storms, and locate dwellings accordingly. A lightning rod may be useful. Hills, trees, and other natural windbreaks may offer protection against the heavy winds that may accompany such storms. Do not, however, build beneath a tall tree, as lightning tends to strike the highest point.

Floods: Be sure dwellings or future building sites are above possible flood levels.

Snowstorms: In areas of heavy snow, it is important to be prepared for the cold and dampness that go along with the snowmen and ski slopes. Access to and from home for supplies is an important consideration. A dry storage area for supplies will be important as preparation for being "snowed in". If you are well prepared, it may be quite enjoyable, and allow time for constructive meditation.

Hurricanes: Sturdily constructed homes will be an

important asset, as well as storm shutters, windbreaks, and other protection against heavy winds and rain.

Tornadoes: An underground storm shelter and good karma are essential in tornado areas.

Earthquakes: These areas are more geological than climatic, but still fit into the present train of thought. Avoid buying land near an active fault. These faults can be located by checking a geological map of the area. If possible, build houses on solid bedrock. Earthquakes will strain the soil, but not the bedrock itself. Outcroppings are an indication of near-surface bedrock. Dwellings built of wood will "give" more than those made of rock and will not be so heavily damaged.

Dust Storm Areas: Areas that are unfortunate enough to have the combination of sparse vegetation, dry climate and high winds will very likely have dust storms. These can be very damaging to crops, as well as uncomfortable to people, animals, and plants. Windbreaks and cover crops which prevent erosion may be necessary in these areas.

Windy Areas: Areas of high wind can be very damaging to crops. High-wind areas are recognized by wind-sculptured trees, that is, trees with their tops bent or curved downwind.

Fire Hazardous Areas: Some areas become very hot and dry during a period of the year. Fires may be started very easily. In these areas it will be necessary to develop water reservoirs. It will also be important to build a firebreak around dwellings and to clear away all dry brush and branches from around buildings, especially roofs near the stovepipe. Once again, prevention is best, so BE CONSCIOUS.

MICROCLIMATE

A Microclimate exists in a small area where the climate is influenced and usually tempered by various elements of natural terrain. Areas in pockets of mild climate are very desirable. When searching for land, look for these areas, as they may provide an oasis within the desert of extreme seasonal climate.

Direction of Land Slope influences the general climate. Land on southern and eastern slopes catches more sun energy. Northern slopes are cooler. Western slopes are cool in the morning, warm in the evening.

Hills and Windbreaks protect against storms and winds if they are between your land and the direction from which storms come. They may also form rain shadows if they are located in such a way as to catch rainclouds, which will drop much of their moisture on the windward side of the hill.

Rainshadow areas occur where there are tall mountains between an area and the oncoming storms. Clouds are trapped on the other side of the mountain, where they drop their precipitation.

RAINSHADOW

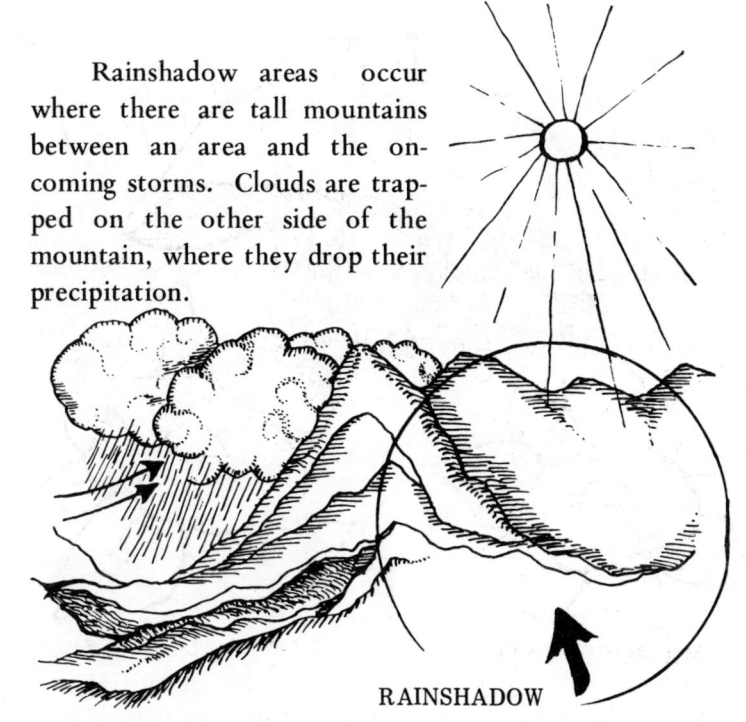

Bodies of Water such as ponds, creeks, rivers, lakes and oceans raise humidity in the area. They lower the temperature during the hot months through moisture evaporation and cooling currents. They also raise the temperature during the cold months through heat absorption during the day and release at night.

Rock Outcroppings, Bluffs and Natural Dikes on a sunny exposure absorb and retain heat during the day and release it at night.

Forests form natural windbreaks and also create centers of high humidity.

It is of great advantage to buy land that features these natural elements of microclimate. Developing microclimates will be more thoroughly discussed in the next chapter.

VEGETATION

By observing the local vegetation you will get a feeling for the land. Much is to be learned about the water, climate and soil of an area by seeing what grows there naturally.

Individual desires, plans and needs will influence the choice of mountain or desert. For instance, if you plan to build a log cabin or mill your own lumber, you will want to find a place with abundant trees. In cold areas, consider the availability of fallen trees for easily gathered firewood. In general, as already discussed in the soil section, the lusher the vegetation, the more fertile the future farmstead. However, all plants are not so desirable. Be aware of allergic reactions to certain plants, as well as the not so positive aspects of living with some of them. Here are a few:

Plants which may irritate hay fever or similar allergies:

acacia	ragweed
black locust tree	tanbark oak tree
milkweed	tumbleweed
rabbitbrush	yarrow

Plants which may irritate skin:

black locust tree (thorns)
cow parsnip
nettles (sting)
poison ivy (itch)
poison oak (itch)
primrose (some species)
thistles (thorns)
wild carrot (burrs)

Plants that may make land clearing and cultivation of gardens difficult, as they tend to "take over":

bindweed
cat brier
chickweed
common burdock
devil's paintbrush
greenbrier
groundsel
hawkweed
Johnson grass
knotweed
man-of-the-earth
milkweed
motherwort
musk mallow
nut grass
pigweed
poison oak & ivy
Russian thistle
silkweed
star cucumber
sticktights
tumbleweed
wild carrot
uncontrolled berry vines

Plants that are poisonous to people or animals:

acorns (raw)
autumn crocus
belladonna
bittersweet
black locust
bog kalmia
buckeye
buttercup
daffodils
darnel
dogbane
dog fennel
foxglove
holly
horsetail (outer coating)
ivy
lobelia
locoweed
lupine
mayweed
monkshood
mistletoe
mountain laurel
mushrooms (some spp.)
nettle
oleander
poinsettia
poison hemlock
poison larkspur
poison sumac
pokeweed
sorrel (quantities)
tansy (quantities
water hemlock
wild onions (quantities)
wild parsley
wisteria
yew tree

COMPATIBILITY OF NEIGHBORS

If the land you are considering is within close range of other people, be sure that these neighbors are of a compatible life style. This is especially important if you are sharing creeks, other water systems, or roads. In order to maintain the purest environment, choose neighbors who are ecology-minded and those who avoid chemical pollutants. If you are a lover of wildlife, it may be difficult to live near people who are hunters. If you are like most people who are moving to the land, you are searching for a better way of life. This new life should not be spoiled by incompatible neighbors.

Extreme care should also be taken when choosing land partners. Aside from being neighbors, these partners will directly affect life's harmony and the land's development. Spend time together on "trial run" camping trips. Discuss plans and feelings about the various aspects of developing the land, such as the number of people the land can support, road use, gardens (separate or communal), division of labor, livestock and pets. Many people, for instance, feel that their country home would not be complete without a dog. If your idea of a "complete country home" is constant barking, chasing away birds, deer, and other wildlife, chasing sheep (which can cause hard feelings with sheep owners, possible fines, and even shooting of the dog), and being surrounded by numerous piles of dog excreta (along with the flies they attract), then a dog is just what you need. Few people take on the responsibility and put out the energy necessary to keep a dog under control. Those who do can generally avoid the above mentioned problems. The limiting of livestock and pets, especially those that affect gardens and other animal life should be thoroughly discussed with the other land partners.

Beware of people who tend to push their ideals too strongly on others, as the object of a community should be working together, not working under the command of another. At the same time, a competent leader can be very beneficial to a community. A leader must be able to guide, not to drive, have great knowledge of the land and its creatures, and must be able to teach instead of criticize. Having a common denominator in a community helps to keep everything running harmoniously, be it the community garden, a spiritual leader, meditation, or just plain respect and acceptance of people for what they are, not what you would like them to be.

LEGAL ASPECTS

The general rule when buying land is to be wary. Know what you want and what you are getting. Do not be rushed into making a decision, and be aware before signing any papers. Real estate agents are trained to be salesmen, to show off the good points of the land and not the bad. Some of them are not at all familiar with the many aspects of developing of land.

Land should be surveyed and all corners and rights

of way should be permanently marked, on the property as well as on the deed, so that they can be easily located. Many times the salesman does not know the true location of the property lines, and you, the buyer, may not be getting such valuable assets as springs, creeks, or trees that you think are included in the property. If the property isn't located on a public road, be sure there is Deeded Access to the property. When buying land that is already developed, be sure to get a bill of sale listing all personal property that is included in the sale. Generally speaking, personal property is considered to be any property not permanently fastened to the land. This may include tools, lumber, etc.

Be sure that all aspects of the sale are in writing so that you know what you are getting and what you are paying. There will be costs other than the price of the property, such as drawing of deeds and other papers, title insurance, commissions, and other fees. Be sure you know what part of these costs you will pay.

It is best to have the sale handled through a title

company because these companies are familiar with proper drawing of land papers and searching of the records for encumbrances that may affect the title.

Make sure that all mineral, water and timber rights go with the land. This is very important and don't let anyone tell you that it isn't. Many people have bought land and developed it into beautiful and valuable property only to have it destroyed a few years later by someone exploiting these resources.

If the land is going to be a family or group venture, you should have a joint tenancy deed. In this type of deed, the property automatically goes to the surviving tenants in the event of a death. This can mean great savings because with a regular deed, the property would go through probate, which is very expensive and time consuming.

Do not be timid about asking to have everything put into writing and clarified to your satisfaction. Very often a written document is the only protection you have and it insures that you will be getting what you believe you are getting.

TERMS USED IN CONJUNCTION WITH LAND TRANSFERS

Deposit: A sum of money paid by the purchaser to the seller to bind the deal and show intention to purchase.

Escrow: Legal papers and money held by a third party, usually a title company, until purchase can be completed.

Bill of Sale: Gives title to personal property.

Township: A unit of land measuring six miles square.

Section: A unit of land measuring one mile square.

Parcel: A piece of land divided away from a larger tract.

Encumbrances: Anything that might affect the title or use of the land.

Restrictions: Limits on the use of the land.

Easement: A right to use property for a road, pipeline, powerline, etc.

Reservations: Part of the property is retained by a former owner, such as water, timber or minerals. Any rights not reserved or excepted are generally considered to go with the property.

Water, Timber, Mineral Rights: The right to use or extract water, timber, or minerals, respectively, from property.

Survey: Measuring and marking the boundaries of land.

Legal Description: Written description of land, including township, section, and lot number.

Grant or Warranty Deed: A written document by which the ownership of land is transferred from one person to another and title is guaranteed by the seller.

Quitclaim Deed: A transfer of land ownership where the title is not guaranteed by the seller.

Trust Deed: A written document by which the title to the land is conveyed, usually for securing a note.

Note: A written document showing evidence of a debt.

Title Insurance: An insured statement of the condition of the title, protecting buyer's rights against claims of others.

Closing of Escrow: When all terms and escrow agreements have been met.

Recording: Putting a deed or other written document on public record, usually at the Courthouse and County Seat.

Mortgage: The pledging of property to a creditor as security for the payment of a debt.

Foreclosure: To deprive of the right to redeem a mortgage when payments have not been met.

Reconveyance: To return a title to a former owner, the title usually being a trust deed.

MEDITATION

Developing a creative, self-sufficient environment can be highly fulfilling. Our home is our center and from here we find the center within our Selves.

When considering a piece of land, take a few quiet moments to observe your inner feelings as they relate to your outer surroundings. Close your eyes and listen to the land, not with your ears but with your heart. This is the supreme test. If the land passes this test, you're ready to sign the papers.

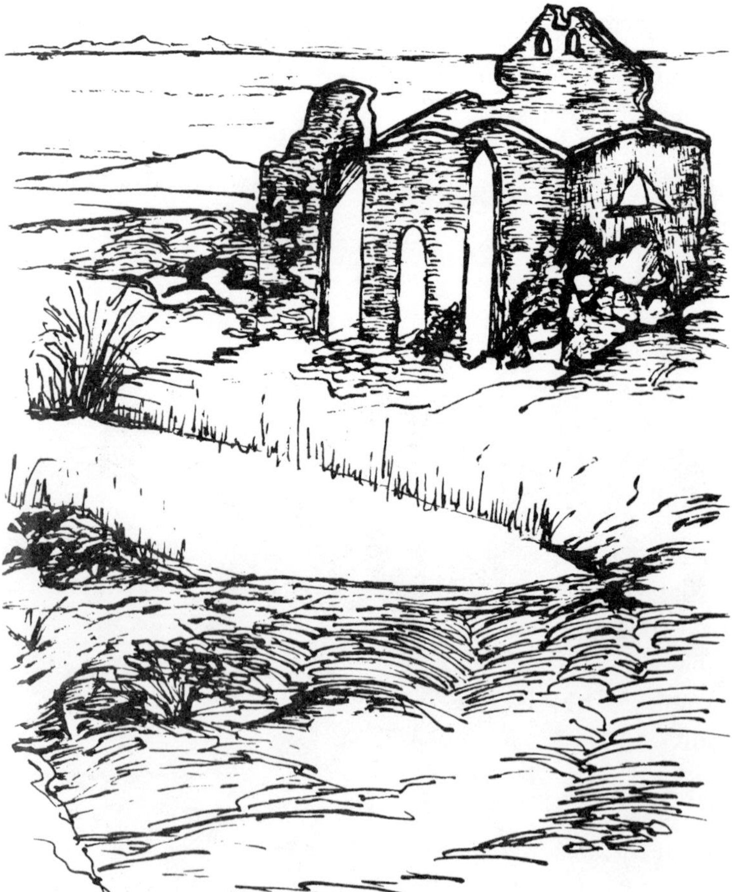

CHAPTER II
DEVELOPMENT OF LAND

Once you have found the land of your dreams, careful development is important to keep it from turning into a nightmare. There are usually initial hassles to overcome in the beginning, but by doing everything completely and consciously the first time, much energy will be saved.

Your needs, plans, finances, and life style will affect how you will develop your land. This book is based on setting up a farmstead. Farming, although not always a necessity, will take you a long way towards self-sufficiency. Gardens, orchards and livestock can provide a hundred percent of your food if planned and managed carefully.

Developing land is like making pottery. You have begun with the raw earth and now the task is to transform it into something useful, something beautiful. Sometimes, through lack of concentration or the wrong formulas in the beginning, it may just become another lump of mud. But through conscious effort, patience, and hard work you will find yourself turning your land (on the wheel of life) into the perfection of form. Glaze it with the glass of creativity and fire it with the Eternal flame and you will manifest perfection of environment.

ROADS

Roads are the first consideration, as they provide a way for you to get to where you're going. A road may be a simple trail or pathway, but if you plan to do much building, a drivable road is very desirable for carrying building supplies. As mentioned before, it is good to buy land with the best possible road already there, or an agreement with the seller to take care of putting one in. If, however, you find yourself without a road, you will have to put one in yourself. This can be to your advantage, as you will be able to (and should, if possible) supervise what is done and have the road put exactly where you want it.

Cutting a road is a job for an experienced equipment operator. He will know where it is possible to put a road and where it would not be possible due to the chances of sliding or erosion of the surrounding terrain. If money is a problem (cat drivers charge $15 to $30 per hour) perhaps you can offer to be an assistant on your road construction. The driver may give you a lower rate, but, even if he doesn't, your help would probably shorten the number of hours of working time considerably. As an assistant, you would probably move large rocks, fallen tree limbs, cut and clear brush, set chokers, and a number of small jobs that might even double the driver's time if he had to do them himself. You would also be gaining valuable knowledge and experience that would help you in maintaining the road. Another way to lower the road building cost is to get people who are sharing the road to help share in the time and expense. Here are the main points to consider in road building:

1. Plan the route that the road is to follow through creeks, around trees, boulders, etc.

2. Bridges across creeks may be necessary. Be sure

they are sturdily built and be especially careful to build them wide enough and above the high water level of the creek.

3. Decide how wide the road will be. Usually, the wider the better (within reason). Width will be dependent on the number of travelers on the road. One lane is usually sufficient, but be sure to allow wide areas for passing and for blind curves.

4. If a road will be a through-road used by several people, don't allow it to cut your property in half. It is not always desirable to have people driving through your land continually. It also means more livestock fencing and may make water or fuel lines more expensive to install.

5. It is good to have the road built along the edge of a forest so as not to divide meadowland in half.

6. Do not, however, build road too close to trees so as to injure them. Road should not extend within the edge of the widest limbs and branches of the tree (drip line).

7. Allow for places to turn around, especially near your house. Also be sure to allow for passing space on the road if it is narrow.

8. It is best to cut a road at the end of a rainy season, while the soil is still moist. This moistness will help the road to pack better.

9. If possible, locate the road on high ground. High ground is generally more stable than lowlands and there is less chance of flooding and washing out. Drainage on high ground is also much better.

10. Have road cut as level as possible. Steep roads wash out much more easily than level or gently

sloping land. Cross-slope, rolling land is good. This is where there are several humps in the road leading upward. It allows for drainage across rather than down the middle of the road.

11. A drainage ditch on each side of the road lengthwise is essential to let water run off. It should be deep enough to catch the water flow, but not deep enough to erode the hillside. Also, by placing medium-sized rocks in the ditch at intervals, the water flow will be slowed down enough to prevent the ditch from eroding.

12. On a steep road it is important to divert the water from draining down the middle of the road or cutting across the road. To prevent this, cut road surface at a slight incline from the center of the road to the sides, with a greater incline towards the drainage ditch running along the side of the mountain. Where road runs through places where there is level land on both sides of the road, incline may be equal on both sides (crown).

13. Culverts should be placed wherever water drainway crosses the road. This may occur at low points in the road or at places where water washes down from the hillside.

14. On long stretches of road, it is good to put in a culvert even when there are no water drainways washing down on the road. The heavier the rainfall in an area, the closer together you will want to space culverts.

15. Seed the sides of the road with a good grass seed, such as rye or fescue. If road is not to be graveled or topped immediately, the entire surface of the road may be seeded.

16. In clay-soil areas, a layer of sand placed on the

road before topping it (with gravel, etc.) will greatly improve the drainage and stability of the road. To form a hard surface, a mixture of gravel and lime will form a sort of pavement. This is mainly suitable in areas where there are natural lime deposits or for short stretches of road, as lime may be costly to buy.

17. Graveling the road should be done during the wet season so that the gravel will pack into the road well and hold the road together. Graveling is generally done with a dump truck which can spread the gravel to the desired thickness, or it can be spread with a shovel from the back of a pickup truck. Be careful not to fill the drainage ditch with gravel.

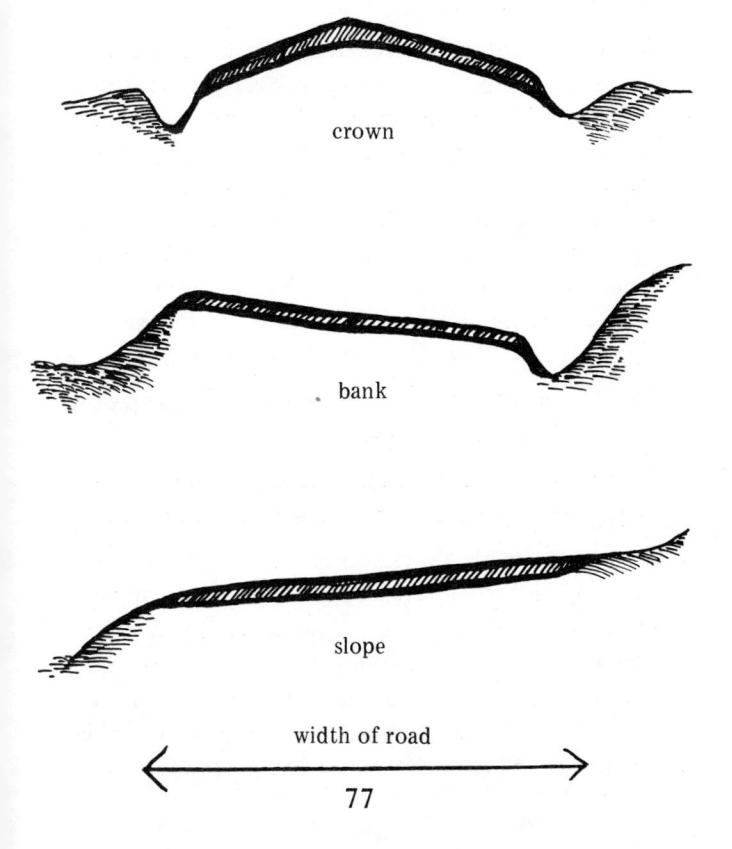

crown

bank

slope

width of road

UPKEEP AND USE OF THE ROAD

1. Avoid driving on the road just after the first heavy rain of the season. The dust formed on the surface of the road during the dry season turns to heavy mud.
2. If a road begins forming ruts, try not to drive in them, but on the sides of them. Driving in them makes them deeper.
3. It is good to gravel the road at least once a year during the first few years to keep the road in good shape. At the beginning of the wet season would be a good time to do it.
4. Keep a pile of gravel on the roadside to fill in holes as they develop. This is known as preventative maintenance.
5. Keep a lookout for erosion spots during the wet season and dig ditches, use gravel, or put in culverts (whichever is appropriate) before these spots become a major problem. To divert water from running down the center of the road, cut small V ditches at a downward angle from erosion flow to drainage ditch. It is best to do this work during a rain because at this time water erosion flows are more obvious.
6. General Rule: If you are coming downhill and you meet someone on the road coming uphill, pull over and let him pass.
7. Drive slowly, especially past other people's houses and gardens, and at curved places in the road.
8. If you drive in the same tracks each time you travel a dirt road, the tracks will eventually become ruts. When it rains, water will run down the ruts, causing the road to wash out. If you drive in a different track each time you travel the road, this will not happen and the road will become packed over the entire width of the surface.

9. Be kind to the roads you use, even if they aren't your own.

CHOOSING A HOUSESITE

Careful planning is very important when choosing a place for the housesite, as well as for the garden, greenhouse, barn or animal dwellings, and various other outbuildings. As in choosing the best land to buy, but on a smaller scale, the same important considerations apply:

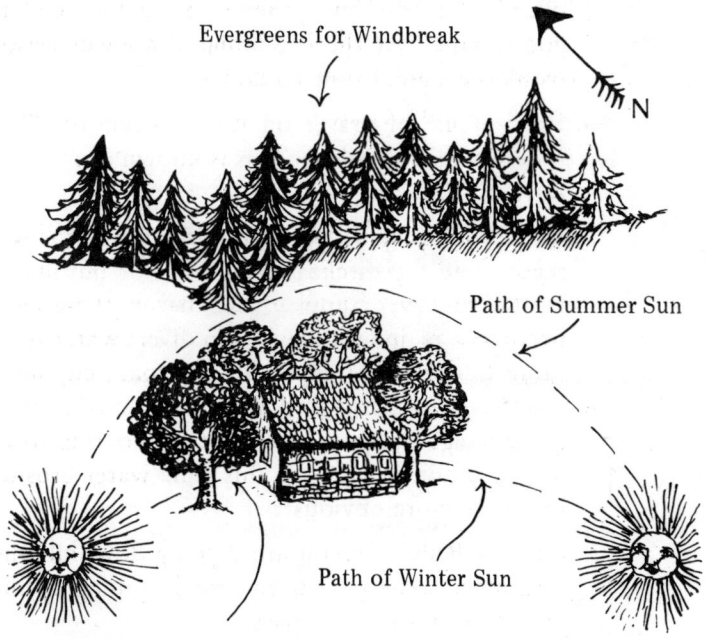

Evergreens for Windbreak

Path of Summer Sun

Path of Winter Sun

Deciduous Trees shade house in summer and lose leaves in winter, allowing sun to reach house.

Place house where it will be shaded in the summer, allow for full sun in the winter, and provide for windbreak during the stormy season.

LOCATION AWAY FROM NOISE, NOSE, AND SIGHT POLLUTION

By spending time on your land, you will find certain areas that are quieter than others. Generally, the lower you are in a valley, the quieter it is, as hills and trees help to make an area cushioned against outside sounds. You will also find that certain areas are more picturesque, more pleasing in vibrations. These places are often found high, overlooking the valleys. Arrange the various buildings taking these points into consideration. Try to arrange them so that there is no interference from one building to another and so that energy is not wasted. Careful thought is important to creating a harmonious atmosphere. Here are a few examples:

Don't put a power-tool workshop near meditation room.
Don't put a power-tool workshop near the garden.
Don't put garden or house too near the road.
Don't put greenhouses underneath trees.
Don't put animal dwellings upwind from house.
Don't put outhouse too near house or water lines.
Don't put garden shed far away from garden.
Don't put outhouse too far from home.
Don't put pantryhouse too far from kitchen.
Don't put compost bin upwind from house.
Don't plan for refrigeration or ice room to be in a hot, sunny spot.
Don't put woodshed too far from house.
Don't put garden too far from house.
Don't spread farmstead out too much. By keeping everything nearby, much energy is saved in going from one to another. Energy and expense will also be saved in developing water and possible electrical lines. It is also good to locate home overlooking other buildings so that farm activities can be observed from the house.

AVAILABILITY OF WATER

Locate buildings that will have running water as near as possible to the water source. It is advantageous to plan for gravity flow of water to buildings, especially house, garden, and animal dwellings.

On the other hand, choose a spot that is on high ground, not in a floodwater path or too near a creek or other body of water. The water vegetation mentioned in Chapter I is also an indication of where you may find winter flood areas.

SLOPE

Locate the house to face south if at all possible, with shade trees on both east and west sides, to provide shade during the summer when the sun is high, and sunshine during the winter when the sun is low. Garden also should be on the south slope. Bedrooms facing east are good, as they catch the early morning light. Consider where the sun rises and sets at the various times of the year.

It is best to build on level or gently sloping terrain, as it may be difficult and expensive to build on extremely steep land.

SOIL

If possible, the house foundation should be on bedrock or on a good sandy or rocky soil. If you are in a heavy clay area, it is best to put down a layer of sand and gravel before building. Garden should be put in a fertile meadow, where soil is of as even a texture as possible.

ROADS

Locate near enough to road so as to be convenient, but far enough away so as to avoid dust, noise, carbon monoxide, and other disturbances.

CLIMATE AND MICROCLIMATE

If you are in an area where climate is not balmy all year 'round, try to locate houses and garden in natural microclimate areas. Shelter and shade are provided by trees, which make excellent windbreaks on the north side of a house or garden, as well as providing cooling shade in the summer. Hills also provide protection from storms. Bodies of water and large rocks help make climate more temperate, and are good to have near house and garden.

LAND STABILITY

Land that is unstable will generally have a rough, slumpy appearance and there may be surface cracks, erosion gullies, or slides. Sometimes the unstable material can be removed, leaving a solid base on which to build. If the base is not of some solid material, it is best to pour a concrete slab to build on. It will be necessary to build a perimeter foundation, which should be of concrete.

VEGETATION

It is good to arrange the house where there is the best combination of trees for shade and windbreak, as well as open meadow for housesite and gardens. If clearing the land is necessary, brush can be used to fill in low places, stop and prevent erosion; and trees may be used for building. Vegetation adds much to the beauty of land, and it is good to leave as much of it as possible in its natural state.

NEIGHBORS

When planning to build, consider the effect of, as well as your effect upon, your neighbors. If four people, each owning a thousand acres, build within a few yards of each other in juxtaposing corners, they might find themselves a bit too close for comfortable living. On the other hand, if your nearest neighbor is very far away, you may get lonely. If you are developing a community with several people involved, careful planning will be important.

LEGAL ASPECTS

Building codes were originally created for the safety and benefit of the people. Many of them have become unnecessary hindrances wrapped up in the "red tape" of bureaucracy. Your lifestyle, the area you are living in, what you plan to do with the land, and your land's visibility and accessibility will be a main deciding factor as to whether or not you will build a house or houses up to code. If you plan to, find out the rules and regulations before you begin to build.

Consider your inner feelings as you examine each possible building site. As mentioned in the first chapter, our home is our center, and the energy we receive from our immediate environment may either encourage or hamper our creative spirit. Choose the building site carefully, considering the subtle vibrations which dwell there.

Use your eyes to look inward, within your own spirit as well as within the spirit of the land on which you will build your home.

Use your eyes also to look outward toward the views and vegetation that are to become a part of your life. Consider the sun and moon, their places of rising and setting, that you might harmonize and receive inspiration not only from the nature on earth, but also from that of the heavens.

CLEARING THE LAND

If the land you plan to use for buildings, cultivation, or pasture is wooded or covered with brush, it will need to be cleared. This is best done with a bulldozer equipped with a brush rake. This method removes the tree and plant roots without removing much of the topsoil.

It is difficult to clear by hand, especially when you want to remove trees, but for small areas it would be feasible. To remove small trees, brush, or small tree roots, a winch on a truck or tractor is useful. Trees that must be removed or those that are already fallen may be used for building with, or for firewood. A chainmill is useful in cutting up logs into lumber and there may be someone locally who owns one and would help do the work. A chainsaw is very helpful for cutting firewood. Brush that you clear may be used for making furniture if the trunk and limbs are of a good size, weaving baskets, or kindling wood. Be cautious, however, when clearing such brush as poison oak. Its roots go so deep that it will come right back unless it is burned out. However, the smoke from it is toxic to breathe. Probably the best way to do it is to dig out as much of the main bush as possible with the bulldozer or by hand, place some dry sticks and hay on the roots, light it, and move quickly to an area opposite the direction the wind is blowing. This should be done during the dry time of year. Or you can just try to live with (and around) it and you will probably build up an immunity to it. Cleared brush and logs may also be used to reinforce the banks of a creek or to fill in low spots, alternating layers of brush with layers of earth.

When clearing land for building sites, trees should be left that will furnish windbreaks and summer shade but allow for the winter sun to reach the buildings. It is important, however, to clear all brush away from around the future building sites, as it may be a fire hazard. All trees should be cleared from future cultivation areas, such as garden, orchard or pasture. However, it is a good idea to leave a strip of uncleared land at least ten feet wide at a right angle to the prevailing wind. This forms a natural windbreak, which is important for protection, especially in windy areas.

Consider the natural landscaping of the area and try not to remove more than necessary, as nature has set up a delicate balance. For instance, removing trees or brush from a slope may cause erosion that would be unsightly as well as endangering buildings nearby. Forest areas provide homes for wild animals and prevent nearby soil from drying out. Be conscious and careful of what you are doing before attempting to alter your surroundings.

LEVELING HILLS AND TERRACING

Sometimes it is necessary or desirable to level hills or terrace the hillsides. Leveling is generally done to even-out a building site, when building a road, or when preparing a gardensite. Level land is much easier to work with and topsoil nutrients stay in place much better.

Depending on the amount of space to be leveled, the appropriate method should be used. A rake, shovel, plow, tractor, or bulldozer may be used, depending on the size of the job. The first step is to stake out the area to be leveled. Remove the first layer of topsoil (valuable stuff) and store it in a pile nearby so that later it may be spread back over the leveled area. This is especially important for the garden. A builder's level should be used for leveling large areas because it is hard to tell by eye when an area is level, especially in hilly country. By moving the earth from the high places to the low places, leveling is accomplished.

Terracing is like cutting steps into the side of a hill to provide more usable land and a more gradual slope upwards. The size of each step is determined by the purpose intended. Usually terracing is done to provide garden space, though it opens up a wide range of possibilities for building houses.

Terracing is a means of preventing the topsoil from washing down the hill and also of preventing other forms of erosion. It provides for much better and more economical use of space, especially on steep land. A terraced garden is both beautiful and a means of conserving water and irrigation energy.

When preparing a terraced garden the first step, once again, is to mark out where the various levels are to be, the width of each level, and the depth of slope. Be sure to allow for walkways, preferably on the inside of the terrace (next to hill) so that you won't disturb and erode the earth by walking on the outside edge. Remove the topsoil and save it for spreading back on the terrace levels. Cut and level each terrace using shovel and rake, plow, or bulldozer, depending on the size of the job.

Slope each level at a slight incline downward toward the inside of the hill. This allows rain and irrigation water to wash in toward the hill, rather than washing out over the edges, eroding them and washing them away. It also conserves irrigation water. It is sometimes wise to build retaining walls along the outer edge of each terrace to aid in preventing erosion. These may be built out of stone, wood, tin cans, or whatever happens to need recycling. The terrace may also be seeded to protect it against the first rains. If the terrace is to be used as a garden, it may be seeded with a cover crop which will add valuable nutrients to the soil while it prevents erosion.

BUILDING HOUSES

Houses, barns and other dwellings should be built to reflect the personalities of the dwellers, be they people, chickens, cows, or tools. They must also be built to suit the purpose intended. The following is a list of possible structures for a complete, self-sufficient homestead:

Home	Garden Shed
Greenhouse	Outdoor Privy
Chickenhouse	Sauna
Goat Shed	Pantryhouse
Barn	Woodworking Shop

The dwellings should also be built to reflect the personality of the surrounding land, so that it dwells in harmony with nature and not as an obstruction to it.

SELECTION OF MATERIALS

Materials should be selected to suit the individual taste. Here are some of the advantages and disadvantages of some materials:

Wood: This is perhaps the most popular of all building materials. Wood has warmth and a feeling not found in other materials. Wood is light, strong, beautiful, is a good insulator, and is easy to work with.

Logs: These are good if they are readily available on your property. Log construction is inexpensive, beautiful, and strong. Be sure that logs are of a type of wood that is durable. Many hard woods rot quickly when they weather.

Stone: This is fireproof and may be readily available on your property. Stone is a poor insulator. It is easy to work with. In earthquake country, it is advisable not to build stone walls too high.

Reinforced Concrete: It is strong, inexpensive, fireproof, but also a poor insulator. It is the best material for a foundation because it does not rot and insects will not attack it.

Adobe Bricks: These are cheap, fireproof, and a good insulator. This type of construction is suited to dry areas, though it would be suitable in wet areas if it is waterproofed or provided with a protective porch roof to shield the walls from rain.

FOUNDATION WITH BASEMENT

A basement-type foundation is the most practical foundation recommended for farm house construction. This type of construction gives the storage space and extra room that is often necessary. It also makes the plumbing and electrical systems accessible. This alone is worth the additional cost of basement-type construction.

Reinforced concrete is the best type of construction for a basement. It is best that the house be located on

FOUNDATION (Top View)

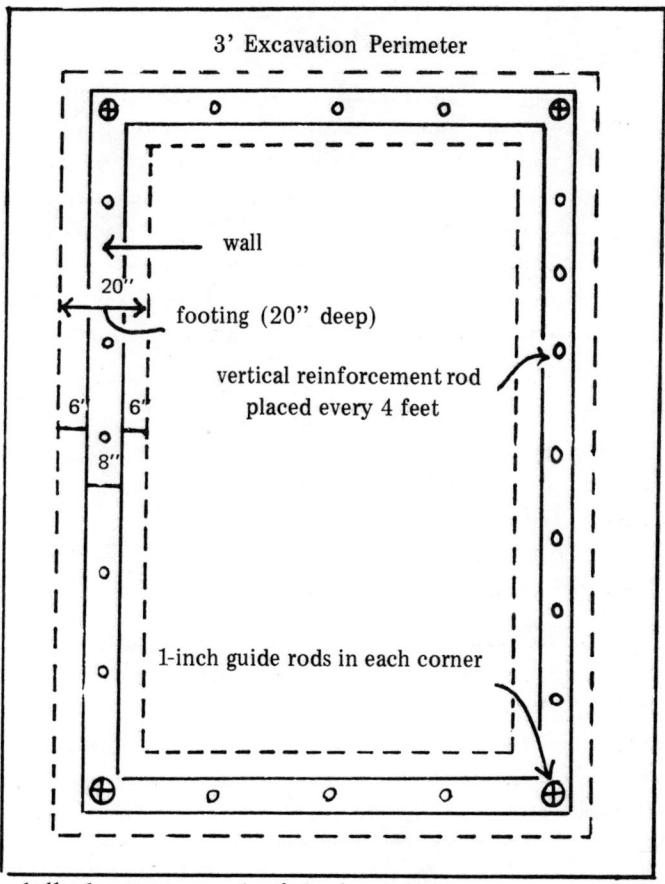

3' Excavation Perimeter

wall

footing (20" deep)

vertical reinforcement rod placed every 4 feet

1-inch guide rods in each corner

a hillside so one end of the basement can open out on a lower level. This will prevent the basement from flooding and will make it easier to move things in and out of the basement.

In areas of severe winters, a deep foundation is necessary to prevent frost damage to the concrete, caused by freezing of the earth.

The temperature of a basement remains fairly constant, which makes it an excellent place to store food. The basement is also an excellent place to build a laundry room, install a wood furnace, workroom, root-cellar, storage room, water heater, and to store firewood. A basement helps to keep the house warm and dry and allows inspection of the underside of the house.

The first step in the construction of the basement is the excavation. This can be done best with a bulldozer or a front-end-loader but, if this equipment is not available, other equipment may be used or the excavating may be done by hand. Excavation will need to be about 7 feet deep, but will vary with different plans or building codes. The excavation will extend 3 feet beyond the perimeter of the outside edge of the footing trench.

After the excavation is completed, dig the footing trench within the excavation and 20 inches below the floor level of the excavation. The footing trench should be 20 inches wide and should extend about 6 inches beyond the perimeter of the wall of the building. It should be centered under all load-bearing walls. A load-bearing wall is one which supports the roof or some other load.

Since the basement is the foundation of the house, care should be taken to make it as square and level as possible. A good way to check the foundation forms for square is to measure diagonally across the corners of the forms, adjusting the forms until the diagonal measurement is the same for all corners.

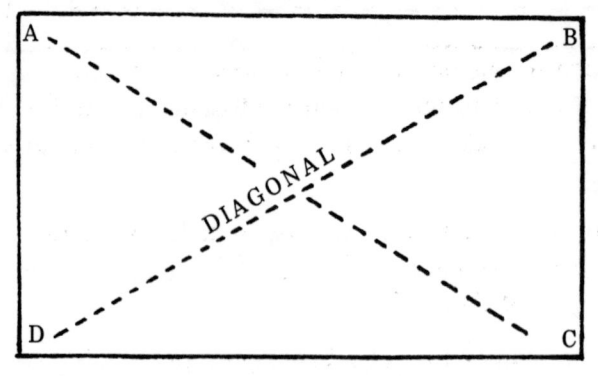

If the measurement of diagonal A-C equals the measurement of diagonal B-D, the form is square.

A good way to keep the wall plumb (straight up and down vertically) and the building square is to set a steel guide rod in the center of each corner of the footing where the wall rises, to act as a guide. These rods should be about one inch in diameter and must be straight. They should be long enough to almost reach the top of the finished basement wall. The rods will act as a guide as the form is progressively moved upward. The rods will also make the walls much stronger, as they will add to the reinforcement of the concrete. The rods are inserted into concrete before the footing trench is poured, and allowed to set.

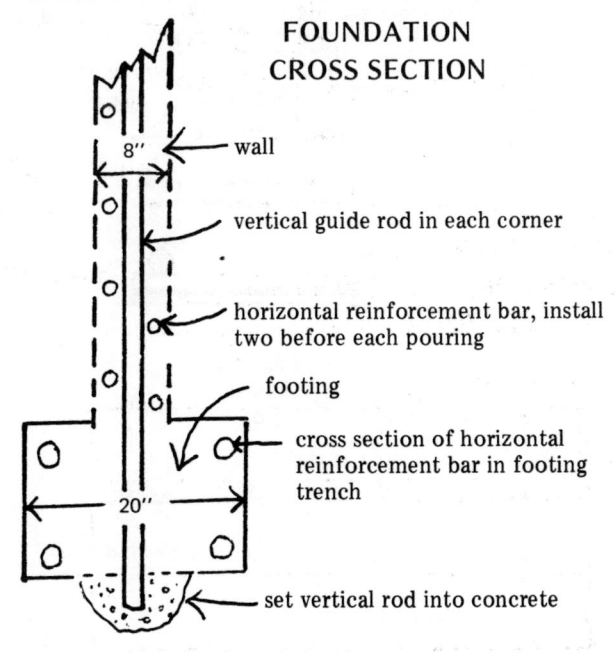

FOUNDATION CROSS SECTION

- 8" wall
- vertical guide rod in each corner
- horizontal reinforcement bar, install two before each pouring
- footing
- cross section of horizontal reinforcement bar in footing trench
- 20"
- set vertical rod into concrete

Between the guide rods, place steel reinforcement rod vertically every 4 feet along the perimeter. Place small rocks under the bottom ends of the vertical reinforcement rods to hold them off the bottom of the trench.

The guide rods and vertical reinforcement rods act as a support for the horizontal reinforcing steel. Short pieces of reinforcing steel are tack-welded or tied with wire to the guide rods at a right angle to the trench to support the horizontal reinforcing steel. This same method is also used to tie the vertical and horizontal rods in the walls together and to support them in the wall forms while the concrete is being poured.

It will be necessary to slip stakes between the horizontal reinforcing steel and the trench wall to hold the steel in place while the concrete is being poured. Be sure to remove these stakes before concrete sets.

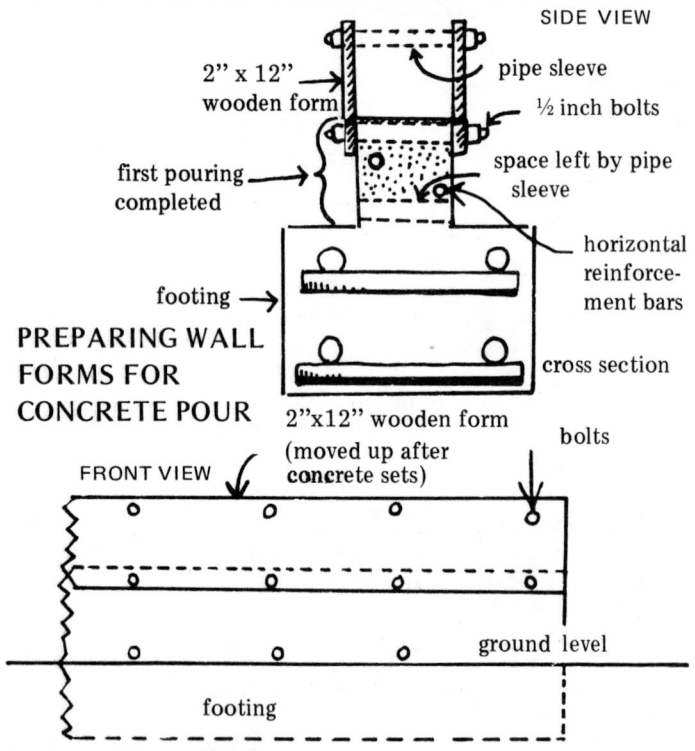

PREPARING WALL FORMS FOR CONCRETE POUR

SIDE VIEW
2" x 12" wooden form
pipe sleeve
½ inch bolts
first pouring completed
space left by pipe sleeve
footing
horizontal reinforcement bars
cross section

FRONT VIEW
2"x12" wooden form (moved up after concrete sets)
bolts
ground level
footing

After the footing trench has been dug, all vertical reinforcement rods placed, and the first four horizontal reinforcing rods placed in the footing trench, the concrete may be poured to fill the footing trench.

After the concrete has set overnight, the wall forms may be set in place. These forms are constructed of 2 x 12 lumber, as anything thinner tends to buckle unless reinforced with 2 x 4 wooden braces. The 2 x 12 wooden forms are constructed as shown in the diagram. The pipe sleeve holding the bolts creates a hole through the concrete wall which then allows the form to be moved up and the bolts to be moved up into place. In this manner, the top sleeve of the first pouring becomes the bottom sleeve of the second pouring, and so on.

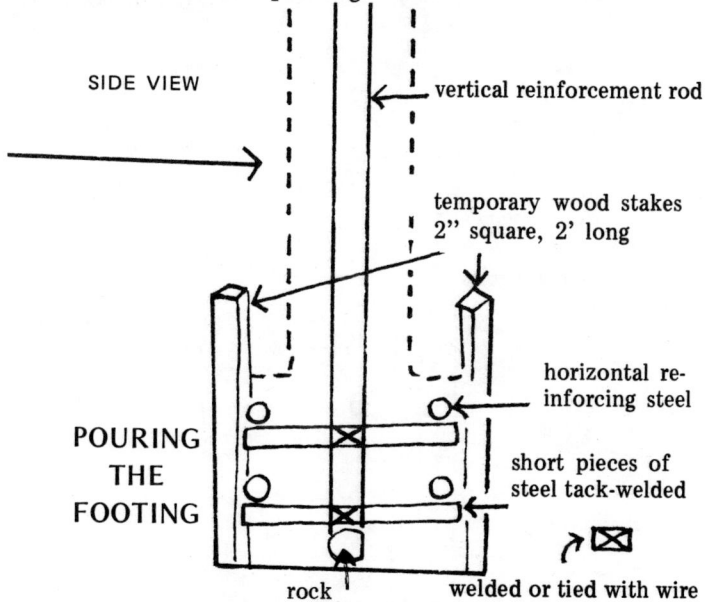

SIDE VIEW

vertical reinforcement rod

temporary wood stakes
2" square, 2' long

horizontal reinforcing steel

POURING THE FOOTING

short pieces of steel tack-welded

rock

welded or tied with wire

When the forms are together for the first pouring, two horizontal reinforcing rods are then installed and the concrete is poured. After this concrete has set overnight, the forms may be raised and the above procedure repeated.

Forms may be continuous around the entire wall or they may be used in short lengths of 4 or 8 feet and the wall poured in block form.

After the basement wall has been poured, install a 4-inch drainpipe around the perimeter of the footing, fill hole around drainpipe with gravel, backfill hole around basement wall with sand, and mound earth up against outside basement wall to prevent erosion.

½ inch anchor bolts welded to each vertical reinforcement rod

earth mounded up to prevent erosion

plug holes on the outside wall with concrete

vertical reinforcement rod

ground level

leave holes open on inside wall for hanging shelves

SAND FILL

horizontal reinforcing rod

(cross section)

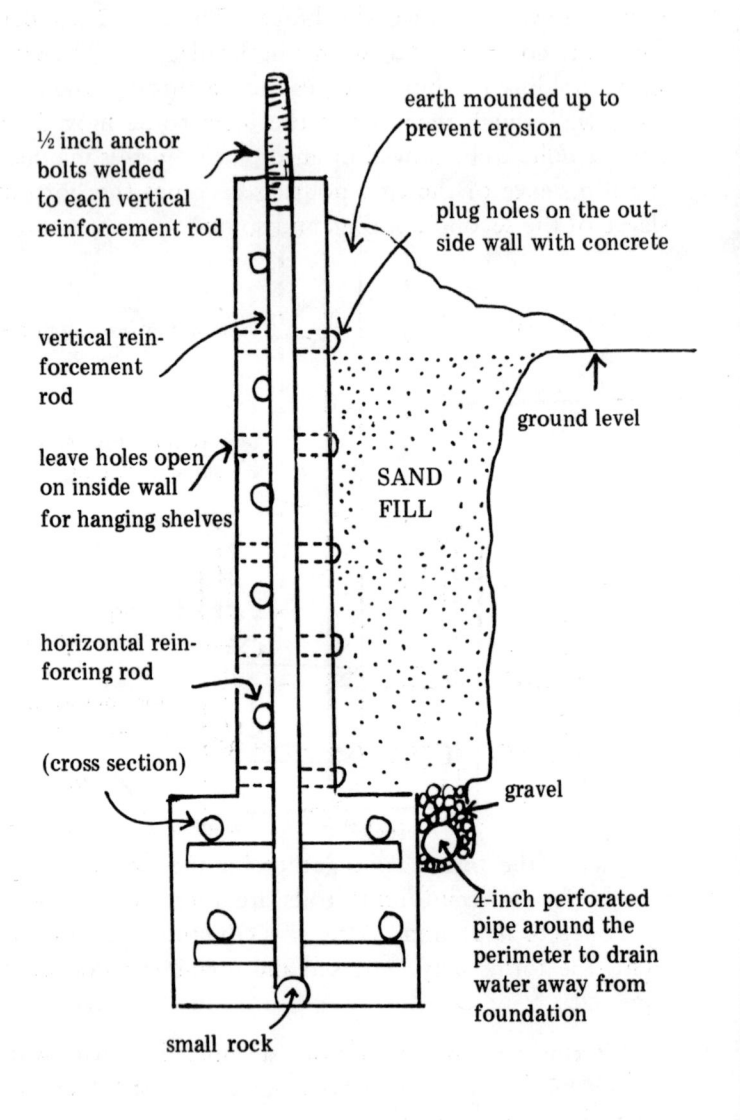

gravel

4-inch perforated pipe around the perimeter to drain water away from foundation

small rock

CROSS SECTION OF COMPLETED WALL

FOUNDATIONS WITHOUT BASEMENT

Solar

For building a solar heated house a shallow basement foundation, about 4 feet deep, is installed in the same way that the basement-type foundation is constructed, as mentioned earlier. This "basement" is dug into a hillside, filled with stones, and a glass window is installed on the south side to catch the sun's heat during winter months. Complete instructions are given later in this chapter.

Perimeter

A perimeter-type foundation is constructed just like the basement-type foundation, except that the basement is excluded, eliminating the need for excavation. A 20-inch-deep footing trench is dug around the perimeter of the future building and filled with concrete. Forms, constructed as before, are built to bring the perimeter 1½ to 2 feet above ground level. Earth should be mounded up against the outside of the wall to prevent erosion around the base of the house. The floor may be of dirt or a concrete slab may be poured at ground level.

Piers

This is the simplest type of foundation. Piers, either purchased or constructed of concrete are placed at approximately 6-foot intervals around the perimeter of the future building. Piers are set on ground level, preferably on top of a layer of sand and gravel or concrete. Sills are then nailed to the tops of the piers, or on a slanting hillside a post is secured from pier to sill and construction of the dwelling is continued.

To make piers:

1. Cut forms from ½ inch exterior grade plywood in the shape of a square 1 foot x 1 foot x 1 foot; or in the shape of a symmetrical trapezoid 8 inches along the top, 12 inches along the bottom, and 12 inches high.
2. Nail four of these together to form a four-sided form, and place where they are to go along the perimeter.
3. Pour concrete, mixed one part cement, two parts sand, three parts small gravel, and enough water to make it workable.
4. A 3½ inch length of 2 x 4 is set into the concrete flush with the top of the pier just after concrete is poured.
5. Keep piers damp for about seven days.
6. Sills are then nailed to the wooden block and construction of the house is continued.

Trapezoid form ready to pour concrete.

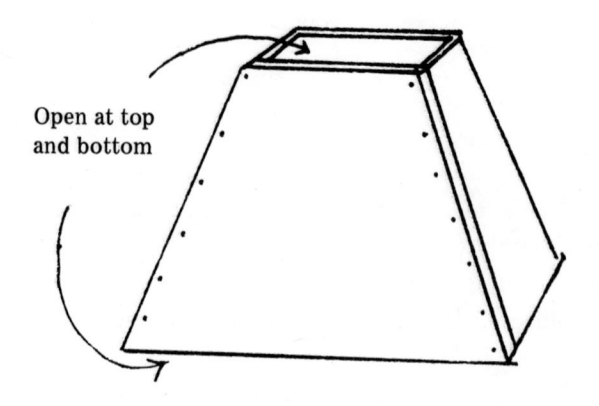

Open at top and bottom

HOUSE CONSTRUCTION ON BASEMENT FOUNDATION

Floor

When the foundation has set for several days, the concrete will be strong enough that the joists may be bolted in place. The joists are beams running parallel across the width of the foundation, on which the floor is nailed. Be sure to bore holes so that joists will line up and be 4 feet apart on centers and fit onto the bolts that

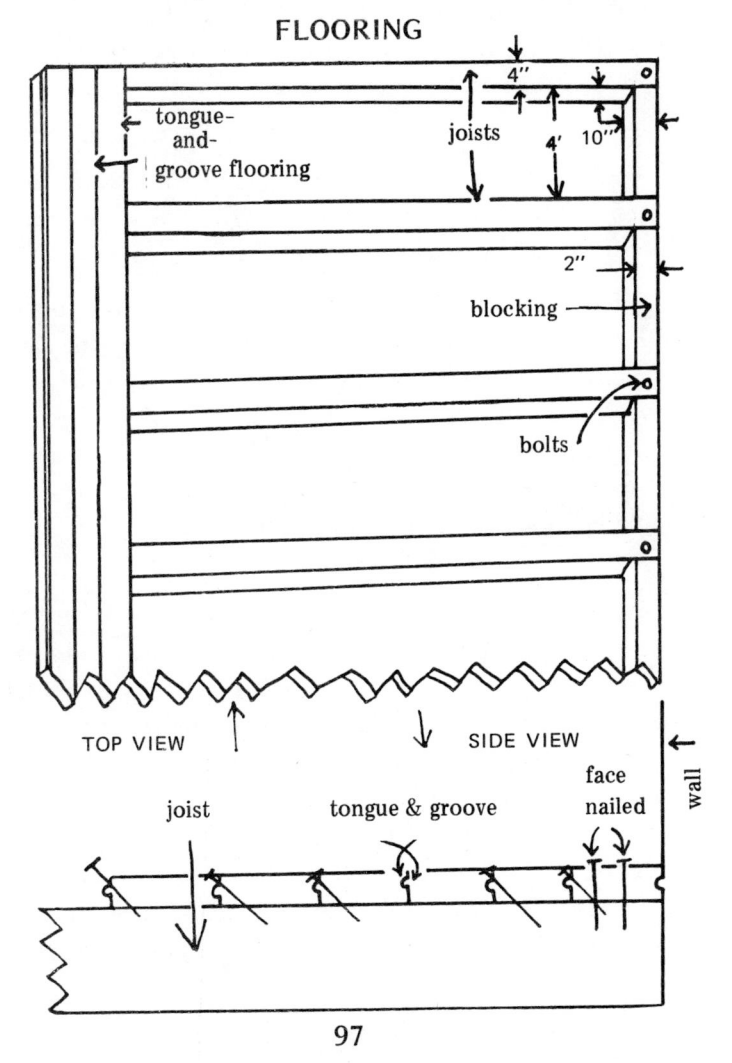

FLOORING

tongue-and-groove flooring

joists 4' 10"

4"

2"

blocking

bolts

TOP VIEW   SIDE VIEW

joist   tongue & groove   face nailed   wall

you have set into your foundation. This can be done by marking off the foundation and measuring to the center of the foundation bolts to see that they are properly located on 4-foot centers. If the bolts are off center, this can be corrected by drilling the holes in the joist to correspond with the misalignment of the bolts.

After the joists are in place and bolted down, 2-inch blocking (the same width as the joists) is fitted and nailed between each joist and the next, flush with the outer ends of the joists. The joists should then be painted with a coat of raw linseed oil to preserve the wood.

After the oil has soaked in, you are ready to lay the floor. The flooring should be of a good grade of dry lumber. If it is not dry, it will shrink, leaving cracks between the boards. Two-by-six tongue-and grove kiln-dried Douglas fir makes excellent flooring.

Face-nail the first piece of flooring to the joists with the groove to the wall. The rest of the flooring can be nailed by starting the nail on top of the tongue and driving it at an angle toward the board. This type of nailing pulls the boards together tightly and also leaves the nail hidden from the finished floor. When all boards are nailed into place, sand the finished floor.

Frame

After the floor has been sanded, you are ready to lay the bottom plate. The bottom plate is generally a sound, straight 2 x 4 and is laid around the perimeter of the floor. After the bottom plate has been nailed in place, posts and beams can be cut and nailed in place, as shown in the diagram. The post-and-beam type of construction is the best way to frame a house. This type of construction is stronger and much faster than the standard 2 x 4 frame construction. With post-and-beam construction, 4 x 4 posts are used on 4-foot centers with beams running lengthwise through the building. The beams are generally 4 x 10 or 4 x 12, but vary with the length of span and the load they are to support.

Temporary braces of 1-inch material are generally used to brace the corners and hold them plumb until permanent braces can be put in. After permanent braces are in, the entire frame should be painted with raw linseed oil.

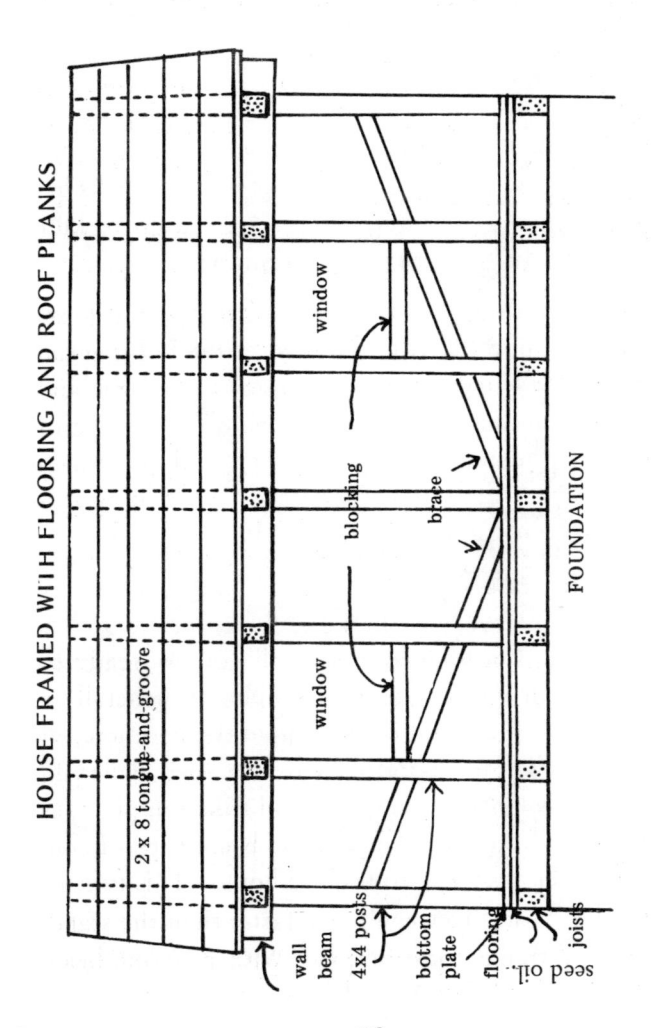

HOUSE FRAMED WITH FLOORING AND ROOF PLANKS

Roof

Laying out the rafters is generally considered to be the most complicated phase of building a house. By following these instructions, it can become a very simple task.

First, find the run of the rafter. This is the distance from the inside of the ridge beam to the outside of the wall beam. Next, determine the rise of the rafter. This will be the run (in feet) x the pitch. The pitch is the rise in inches per foot run.

Pitch of the roof will depend on local conditions. A high pitch is necessary in areas of heavy snow, heavy rain, and in extremely hot areas where house cooling is a consideration (a steep roof will allow more space for hot air to rise within. In cold areas, a less steep roof will be more conservative of heating energy. A 5:12 pitch is a good all around pitch and means a 5-inch rise for each 12-inch run of the rafter. Example: For a run of 10 feet, you would have 10 x 5 = 50, so the rise would be 50 inches.

Cut a straight 1 x 4 the exact length of the run (in this case, 10 feet). Next, cut a straight 1 x 4 the length of the rise plus 12 or more inches, depending on the length of the rafter (in this case, 12 inches would be okay) 50 + 12 = 62. So, in the above example, the second 1 x 4 would be 62 inches long. Cut a plywood brace and guide blocks, and assemble template as shown in the diagram. All rafters should be a near-perfect fit, but it is always best to cut one rafter and try it first.

When rafters are all nailed into place, you are ready to install the roof planks, which run the length of the house, as shown in the diagram.

When the roof is finished, it can be painted with linseed oil and covered with plain felt and shingles.

RAFTER TEMPLATE

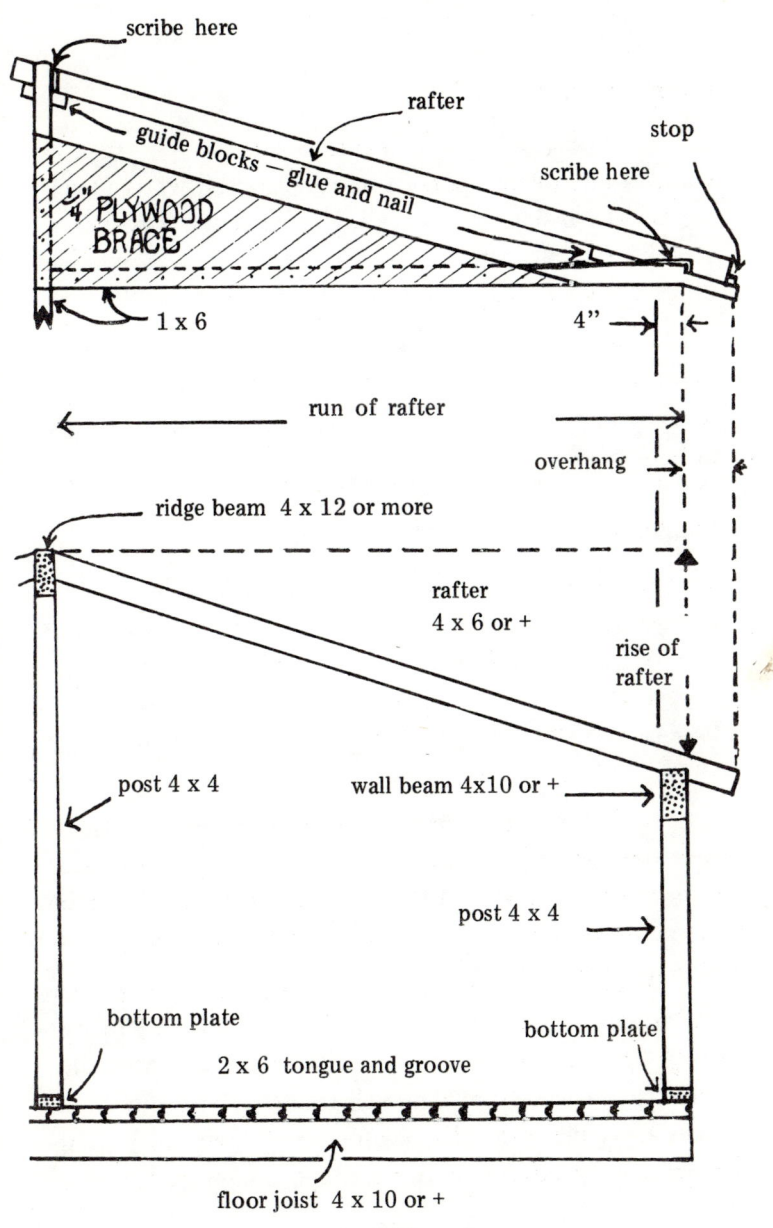

scribe here
rafter
guide blocks — glue and nail
stop
scribe here
½" PLYWOOD BRACE
1 x 6
4"
run of rafter
overhang
ridge beam 4 x 12 or more
rafter 4 x 6 or +
rise of rafter
post 4 x 4
wall beam 4x10 or +
post 4 x 4
bottom plate
bottom plate
2 x 6 tongue and groove
floor joist 4 x 10 or +

Never use asphalt felt (black felt paper) or any other petroleum products when building a house or any other building on your property, as it will pollute the earth when rain washes over it. Raw linseed oil is the best wood preserver you can use and it is nontoxic.

Covering the Frame

This type of house is best finished with a horizontal wood siding on the outside and horizontal wood paneling on the inside. If the siding or paneling is to be installed vertically, it will be necessary to insert wooden blocking horizontally between the posts.

The outside of the house should be painted with raw linseed oil to protect and preserve the wood. It also water-proofs the wood and protects it against attack by insects.

Windows

Windows are an important part of the home. They should be arranged to take advantage of prevailing summer breezes, winter sunshine, nice views, and as an escape exit in case of fire. Be sure that the windows can be easily opened and are low enough that children can climb through them in case of fire.

It is very important to have windows that are tightly sealed when they are closed so that during times of dust storms, heavy smoke, or pollution, they can be kept closed, keeping the air clear inside the house.

When framing the house, be sure to leave the proper size rough opening for the windows. The size of the rough opening is marked on some windows and it is available from the dealers who sell them. There should always be a heavy header beam over the windows. This beam should be continuous through the entire length of the house. A continuous header beam will protect the windows from roof loads and is especially valuable in case of fire, earthquake, or windstorms, protecting the window so that it can be used for an escape exit.

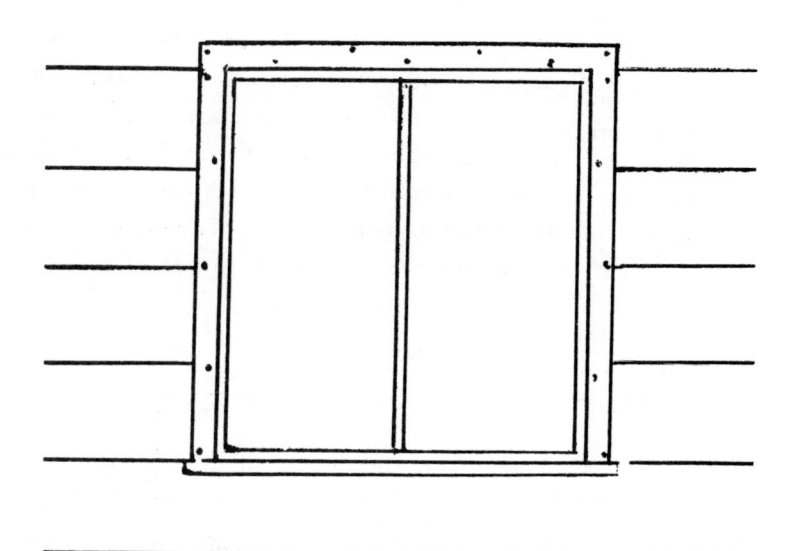

When installing the window, set it in the opening, making sure to divide the clearance spaces on each side so that the window is centered. Level the bottom and tack-nail it to hold it in place.

Plumb the sides and tack nail (tack nailing is driving the nail in part way). Open and close the window, observing the fit on both sides, as well as on the top and bottom, making necessary adjustments by moving the window frame one way or another. Drive tack nails in and complete nailing.

Doors

Doors should be installed to open outward, as this method allows the bottom of the door to extend below the floor line. This type of construction will prevent wind and rain from entering under the door, eliminate the need for a sill (which is a dirt catcher and stumbling block), and is a safety feature in case of fire (an outward opening door allows the occupants to rush the door without the danger of a door jam).

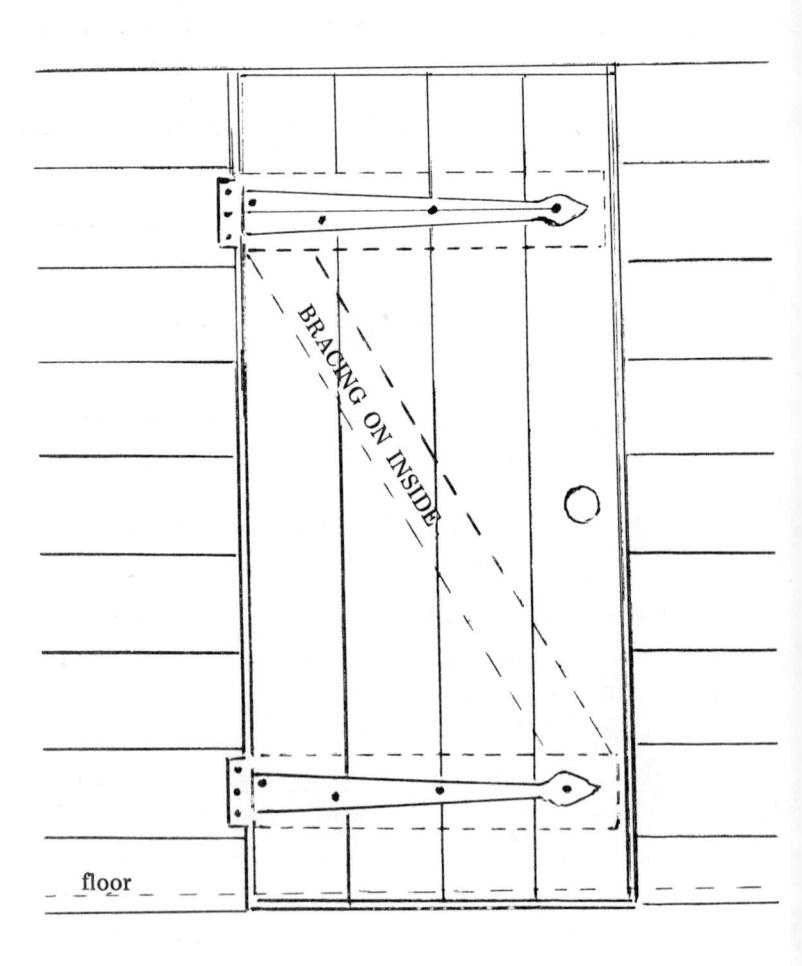

Outdoor Privy

Outdoor privies should be located downwind from the house and at least 100 feet from any well or stream and should be downstream from any source of drinking water. They should be built fly and water tight and have a screened vent pipe.

Be sure to build the privy on stable ground. Dig a hole 3 feet square and about 6 feet deep.

A precast foundation for the privy can be made, or the foundation can be poured in place by covering the excavated square hole with a piece of ¾-inch plywood and cutting the holes for the vent pipe and the seat riser, building the forms, placing reinforcing bars, and pouring the concrete.

Be sure to use a rich mixture of concrete and keep it moist for about a week after it has set.

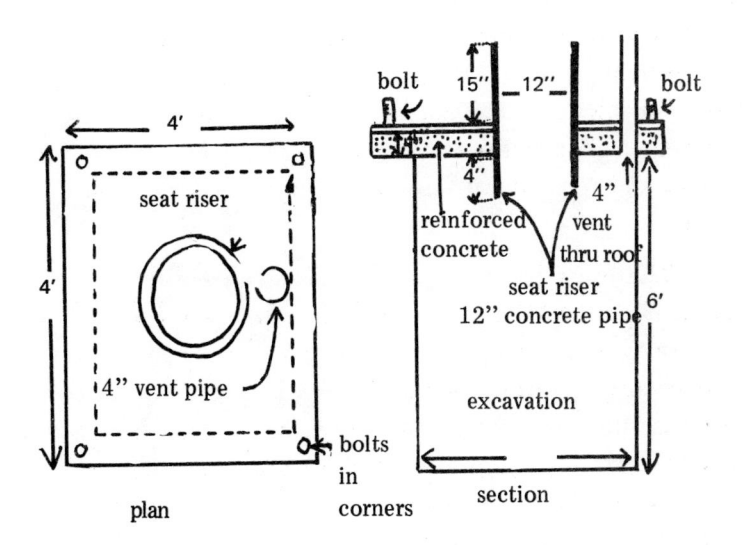

Septic Tank and Drainfield

Septic tank and drainfield must be located at least 100 feet from any stream, well, or any other source of potable water. The location should be carefully selected because it can mean the success or failure of a sewer system. Select a location that is downhill from the house, because the sewer pipe from the house to the septic tank must slope downward (¼ inch to the foot is ideal from house to septic tank). Make sure there is enough room that the drainfield can be expanded later if it becomes necessary. Allow 100 feet of drainfield per person.

Test the drainfield area for porosity by digging several holes about 3 feet deep with a posthole digger. Fill the holes with water and note how long it takes the water to soak into the ground. If all the water is gone within 30 minutes, the site should be considered good. If it takes an hour for the water to soak into the ground, the size of the drainfield area should be doubled. If, after several hours, water is still standing in the holes, the site should be abandoned. If, however, another site is not available, any site can be made usable by removing all of the soil from the drainfield area to a depth of about 3 feet and replacing it with a porous soil, such as sandy loam.

The most basic form of disposal system is a cesspool. Dig a hole 4' x 6' x 6' deep (for the average house). The sides of the hole are lined with wooden planks and the bottom remains earth, with no lining. The water leaches out the bottom and between the side boards, and **bacteria** decompose the solid wastes. A strong, tight cover is placed on top and the sewer drainage pipe from the house is inserted through a hole in the side near the top. This type of system is not recommended and is prohibited in some areas.

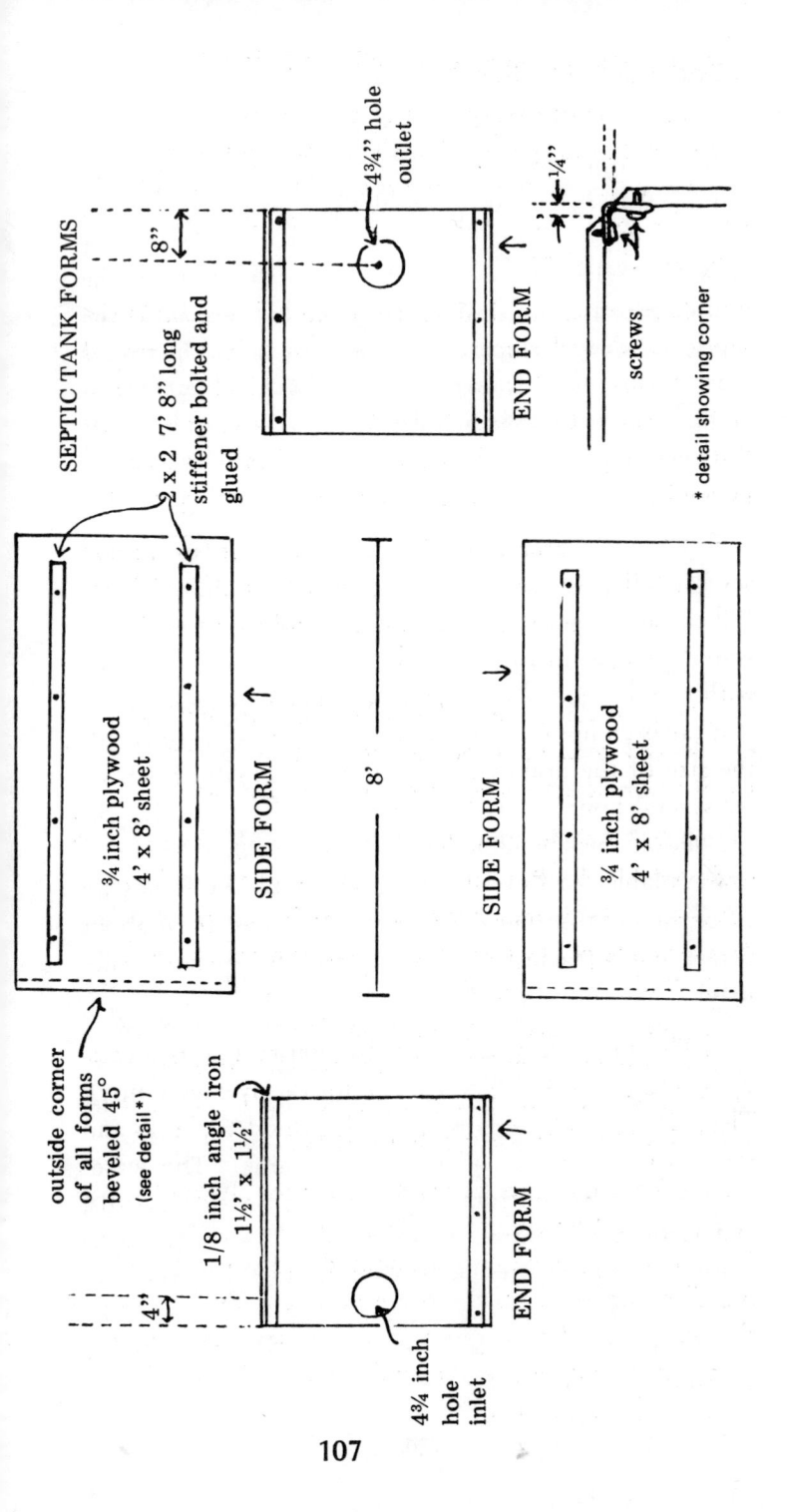

SEPTIC TANK FORMS

A septic tank system is constructed as follows:

After the site has been selected and tests completed, stake out the tank site. For the average house, a tank 4 feet by 8 feet by 4 feet deep is suitable. For a tank this size, the excavation hole should be 4' 8" wide, 8' 8" long, and 5' 6" deep.

Before the hole is dug, it is a good idea to build the forms. This will require two sheets of ¾ inch plywood 4 x 4 feet (for end form); four pieces 1/8 inch angle iron 1½" x 1½" (drill three holes in each one: holes 3/16 inch); four 2 x 2 wood stiffeners 7' 8" long; 16 flathead stove bolts ¼" x 2½"; 12 wood screws 7/8" x no. 12.

Glue and bolt 2 x 2 stiffeners to side forms and screw angle iron corners to end forms, both of which are shown in the diagram. Cut holes in end forms as shown in the diagram.

The hole for the tank can be easily dug with a pick and shovel or a backhoe can be used. If a backhoe is used, the hole will need to be finished by hand.

After hole is completed, assemble the forms and lower into place. Put a piece of 4-inch sewer pipe through the hole in each end of the form, leaving about 4 inches on the inside of the form and 1 foot or more on the outside. The outlet pipe should be 4 inches lower than the inlet pipe. Be sure to plug the outside ends of the pipes so that concrete won't get into them. Reinforcing steel or mesh wire can be put in the space between the form and the dirt wall if desired, but it is not necessary.

Mix the concrete just wet enough that it will handle easily. The drier the mix, the stronger the concrete, but it must be wet enough to be poured and worked so that it will completely fill the space between the forms and the dirt wall. Use a mixture of one part cement, two parts sand, and two parts small gravel. Pour a 2-inch concrete floor on the bottom.

SEPTIC TANK AND DRAINFIELD

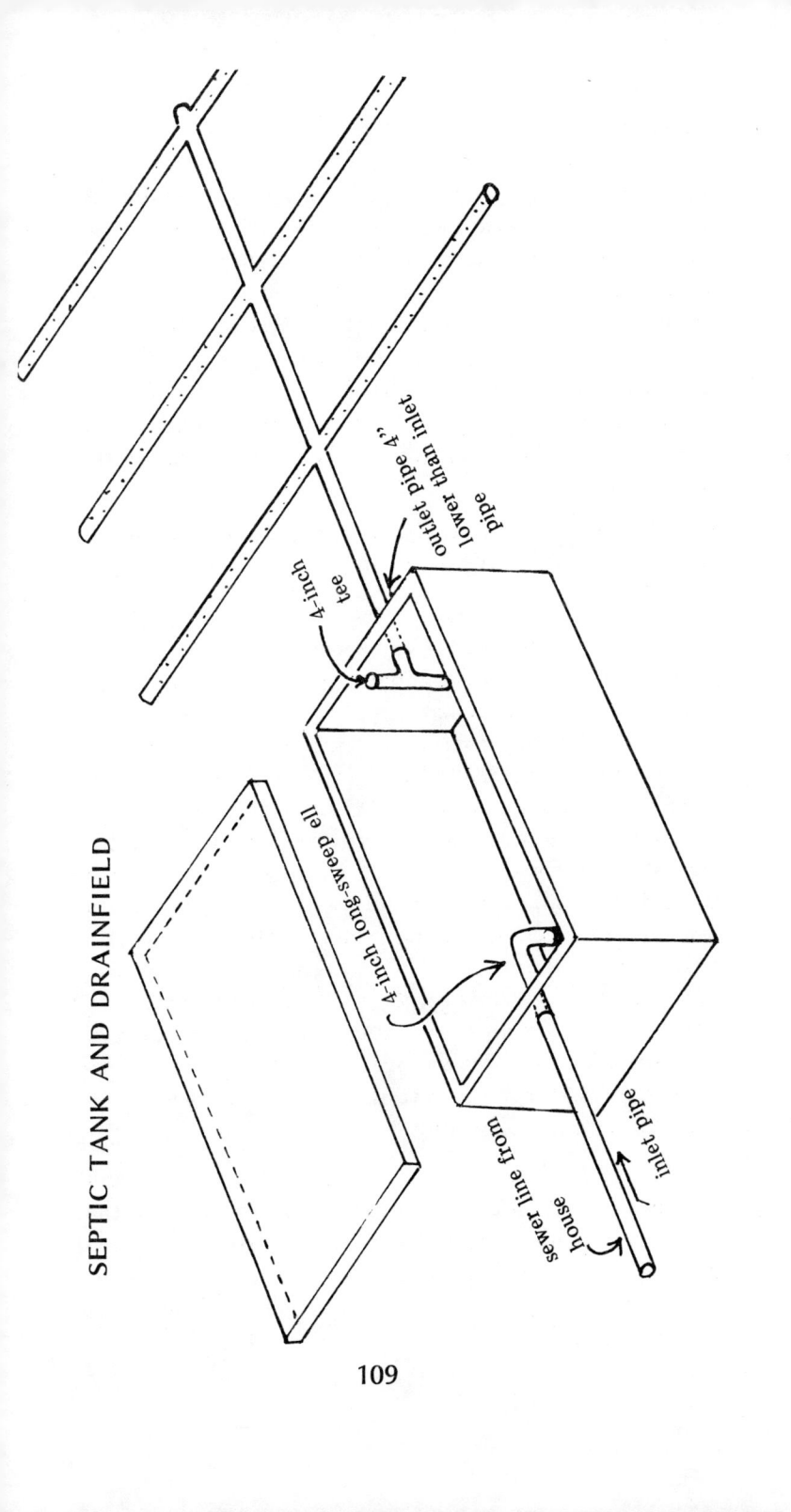

The cover can be made by building a form out of 2 x 3's the size of the tank, 4' 8" x 8' 8" (this should be the inside measurement of the form). Level out a place alongside the tank and set the form against the tank. A piece of heavy reinforcing mesh 4' 7" x 8' 7" should be placed in the form and held up off the ground by several small rocks. The cover can be poured with the same mixture of concrete as the tank. After the cover has been poured and set about 4 hours, cover it with dirt and keep the dirt moist for 10 days. After 10 days, the forms may be removed and the cover turned over on top of the tank. The cover will be very heavy, so be careful when handling it.

Before the cover is put in place, be sure to remove the inside forms and install the fittings on the inlet and outlet pipes.

A septic tank can also be constructed of ferrocement. This is done by lining the excavation with ½-inch-mesh wire and applying a sand-cement plaster. First pour a concrete bottom in the excavation hole. Before the concrete sets place ½-inch reinforcing bars, in a vertical position, about every two feet around the perimeter of the excavation. Leave about a ½-inch clearance between the dirt wall and the rod so mesh can be slipped behind the rod. Line the hole with the ½-inch mesh putting one layer behind the ½-inch reinforcing bars and two layers on the inside. Tie the mesh to the reinforcing bars and tie the layers together. Be sure to allow the concrete in the bottom to set enough so it is not damaged by working on it. Put the inlet and outlet pipes in position (same as for a concrete tank) and you are ready to proceed with the plastering.

The plaster is made by mixing one part cement and two parts sand, adding just enough water to make the plaster workable. Do not allow the finished plaster to dry out. It can be kept moist by spraying with water or by covering with a wet cloth. It should be kept moist for at least one week.

BUILDING FENCES

Fences should be used only where necessary. They are barriers, something unnecessary in a Utopian world, but as we are not there yet these barriers do help keep deer out of the garden, enclose livestock, divide property lines, and are useful any place where you want to keep something in or out.

Fences may be built of various materials. Here are a few:

Hedge: This is a close, continuous planting of the same species of trees or shrubs. They may be either trimmed or left untrimmed, as long as they don't get out of hand. Trimming generally helps them to become thicker. A hedge is probably the most beautiful and natural fence and works well as a divider. Thorny hedges, such as the rose hedge, are good to enclose livestock. The hedges that have berries provide food for the birds and all hedges provide shelter for them as well as for other wild creatures. Quail love to nest in hedges. Hedges are generally maintenance-free, provide a windbreak, and deter wind and water erosion. A hedge for a vegetable garden may not be too practical, however, because it may take it a while to grow tall enough and will shade the garden area too much. Consider how high and wide the particular species you choose will grow before planting it. Some plants used for hedges are: hawthorn, holly, beech trees, box, and roses.

Wood Posts: Combined with chicken wire, this is very suitable for a garden, as the posts may be cut long. Most woods decay when exposed to the weather, especially when buried into the ground. Redwood is perhaps the best and most weather resistant. To protect the posts, they may be painted with raw linseed oil. They may also be protected by charring with a torch the part of the post that goes into the ground plus about six inches above ground. Never use treated wood, as the chemicals used are pollutants to soil and water.

CONCRETE FENCE POST FORM

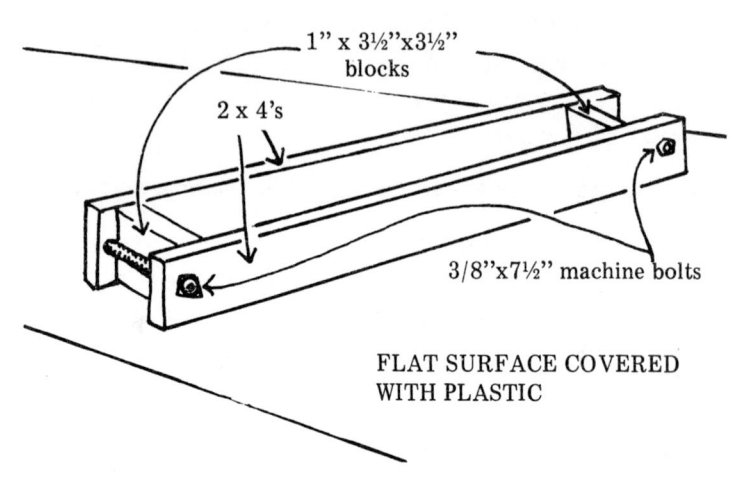

Diagram labels: 1" x 3½"x3½" blocks; 2 x 4's; 3/8"x7½" machine bolts; FLAT SURFACE COVERED WITH PLASTIC

Concrete Posts: This is the most permanent type of fence post. It does not pollute water or soil, does not rot, and it is not affected by insects. They can be made any size or shape. Here is how to make them: Cover a flat area (such as a floor or plywood) with a plastic sheet. Construct forms of the required dimensions with 2 x 4's, as shown in diagram. It is a good idea to make several forms so that several posts may be poured at a time. The forms may be removed the following day after pouring, but the posts themselves should not be moved until they have set for at least two days. Posts should be kept damp for ten days. Fencing is fastened to posts with tie wires, or nails may be set into concrete before it becomes hard.

Metal Posts: These may be used if they are rustproof and unpainted. They should be rustproof because if they aren't, they will rust away in a short time. They should be unpainted because the chemicals from paint which wash into the soil are harmful.

INSTALLING THE FENCE

Install corner posts first. Dig holes about 3 feet deep and set corner posts in concrete. It will be necessary to install braces on the corner posts. These should also be set in concrete.

After the concrete has set one day, stretch a single strand of wire from corner to corner. This wire will act as a guide so that the line of posts will be straight. Make sure that the fence is on your property line. Dig the line post holes about 2 feet deep and 10 feet apart. Set posts in holes and fill holes with sand and small gravel. If you are using metal line posts, holes may not be necessary, as most metal posts can be driven.

You are now ready to stretch the wire mesh or other fencing material you have selected. The fencing should be stretched fairly tight, but if it is installed during warm weather, it will shrink during cold weather and, unless an allowance is made, it may pull out the corner posts or break the wire. After the fence material has been stretched, it may be fastened to the posts.

Other possible fence materials:

stone	bricks	woven tree limbs
logs	water (moat)	climbing vine thickets

MICROCLIMATE DEVELOPMENT

It is possible, through the development of microclimates, to raise the temperature, humidity, and comfort of an area. Ideally, you have already chosen an area that features natural microclimates. By developing these further and by creating new ones, it is possible to increase the productivity, beauty, and fertility of your land.

Hills. When planning out the building sites and garden sites, consider the amount of protection against storms and wind given by the surrounding hills. Hills may be enlarged or constructed, using various fill materials, such as brush and earth cleared from building, garden and pond sites. Nonrecyclable materials that you might normally take to the dump may be used in combination with earth, but be sure not to use anything that may pollute. These "hills" should be packed in as tightly as possible and planted with some sort of shrubbery or grass seed to prevent erosion.

Windbreaks. Windbreaks are important to protect against the force of the wind as well as against the drying effect that the wind has upon plants. Buildings may provide windbreaks for gardens or other buildings when placed between them and oncoming storms. For instance, a barn may become a windbreak for the garden or the chickenhouse. Rock, brick or woven fences will provide windbreaks, as will hedges.

Trees. A line or group of trees provides a windbreak and should be planted on the storm side of the house or garden, taking care not to shade the garden. Trees also raise the humidity of an area, provide shade, give vast quantities of humus for gardening and, if they are fruit trees, will bear fruit. Evergreen trees are good to plant as windbreaks, as they retain their leaves in the winter. Deciduous (leaf-losing) trees make excellent shade trees, as they provide shade in the summer, but lose their leaves in the winter, letting the sun shine in.

Other Plants. Shrubbery entices beneficial birds and animals to your garden. Flowers also attract birds, as well as bees that will help pollinate your garden. All plants give off oxygen, and so increase the oxygen supply of an area. Green plants and grasses that are kept green also raise the humidity of an area. It is good to cultivate various types of plants and shrubbery for these reasons, as well as the obvious reason of creating a beautiful environment.

Bodies of Water. Reservoirs, ponds, swimming holes, fountains, lakes, canals, creeks and springs all help to raise the humidity of an area. They also raise the temperature when the air is cold and lower the temperature when the air is hot. Developing the available water as fully as possible provides plenty of water for all the growing you wish to do. Gardens, animals, trees, green pastures, flowers, and you, all depend on water for your existence. As well as generating life, water may be developed to generate electricity. Its importance cannot be overemphasized and will be covered more thoroughly in the next chapter.

Slope. When planning the various building and garden sites, consider the direction of slope. Here are a few important ones:

barn: east or south slope   orchard: south slope
chickenhouse: east slope   outhouse: east slope
garden: south slope   pantry: north slope
greenhouse: south slope   pasture: west slope
house: east or south slope   trout pond: north slope
workshop: east or south slope

The above list is only a suggestion as to the best place to put the various elements of your farmstead. You will have to modify the layout according to your situation.

Rocks and Stones. Because of their ability to trap the heat of the day and release it at night, rocks and stones are good to have around the farmstead. If there aren't any there naturally, you may have to import them from a nearby creek or riverbed. They are sometimes to be found near natural outcroppings. Medium sized to large stones may be used to provide warmth in many ways:

    Around garden
    Near animal sheds
    Around fruit trees
    Beneath window
    Near ponds
    Around outhouse
    In a rock garden
    Surrounding water lines
    Bordering flowerbeds
    As terrace retaining walls
    Stepping stones through a garden
    As a means of storing solar heat energy for
        warming houses, greenhouses and water

As you gain experience in observing microclimate areas, you will soon be able to discover and develop these areas on your own land.

USING NATURE'S ENERGY

The elements of nature contain a vast storehouse of energy that may be easily harnessed and directed for electricity, heating, and many other uses.

Electrical energy works on the principle of rotation. Natural elements such as water and wind can provide the energy needed to start the wheels rolling so that electrical energy may be generated.

Water falling even from a small spring, can often generate enough electricity for basic needs.

Wind, especially in breezy regions is a good source of power for electrical purposes or for rotating wheels for other purposes.

Solar energy is one of the most powerful sources of energy on the earth. More than 100 trillion horsepower of energy reaches the earth each second, yet only a minute quantity is harnessed for such uses as heating homes, greenhouses or water.

The world is rapidly using and wasting its unreplenishable natural resources. Modern technology, as well as Okie ingenuity, if properly directed could be used to tap nature's storehouse of energy for the benefit of man.

WATER

For centuries, water has been a major energy source for man. Water and waterways have provided a means of transportation and development of cities, countries, and cultures. Water has long been used to provide power for creating mechanical energy to turn wheels. These wheels, then, may be used for such purposes as grinding grain and turning generators.

Generating Electricity

A water wheel is an inexpensive way to generate electricity. Depending upon how much electricity will be used, the water wheel should be built large enough to supply the energy needed. Size of the wheel, as well as the amount of water available, will determine the amount of electricity capable of being generated. If the water supply is small, a water storage pond may be considered as a source of water to keep the wheel rolling. Water flow must be fairly strong and constant to provide the power needed. By locating the water wheel several feet lower than the water source, pressure and flow may be increased. A waterfall is an excellent source to work with, but any reasonably good flow of water will suffice.

The design of the water wheel shown is meant to be basic, to be used as a guide. There can be many variations in design and size, and each water wheel should be designed to correspond with the flow of water available.

The water wheel shown in the diagram is constructed of metal and is an overshot-type water wheel, which

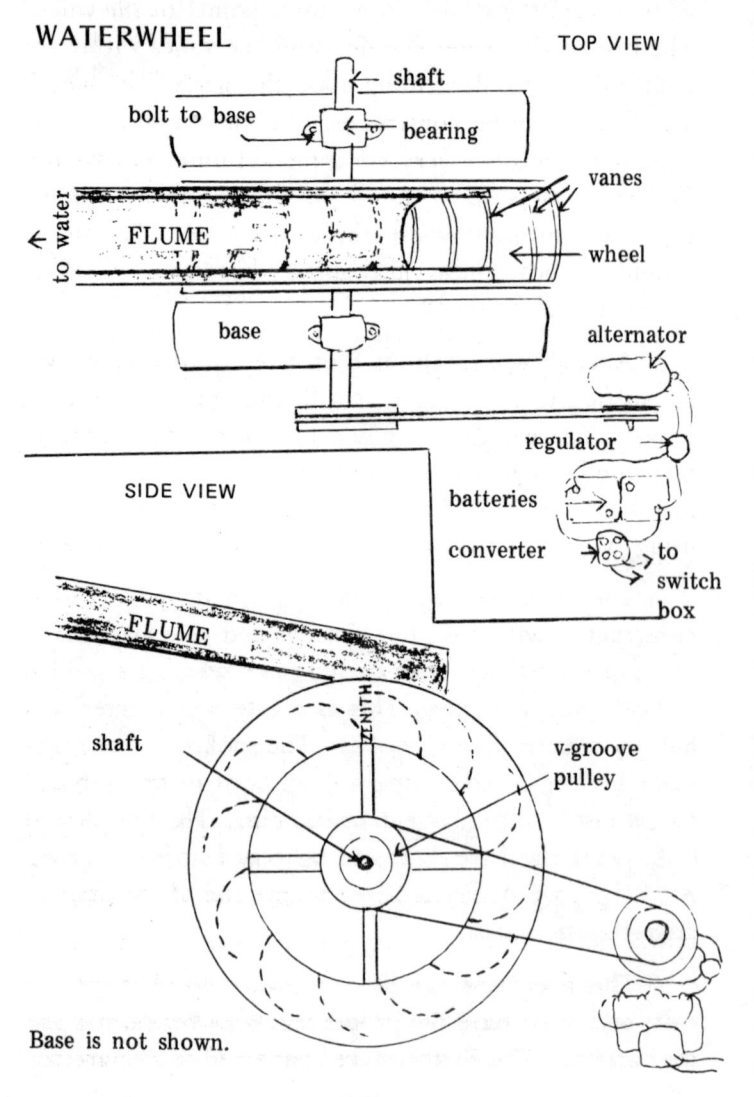

WATERWHEEL　　　　　　　　　　　　TOP VIEW

- shaft
- bolt to base
- bearing
- vanes
- FLUME
- to water
- wheel
- base
- alternator
- regulator
- SIDE VIEW
- batteries
- converter
- to switch box
- FLUME
- shaft
- ZENITH
- v-groove pulley

Base is not shown.

118

means that the water goes in over the top of the wheel. With an undershot-type water wheel, the water goes down under the wheel.

Run a flume from the water source down to the water wheel location. The flume is a trough-like structure running from the water source to the water wheel in such a way that part of the water enters the vanes just as they pass the zenith (top center point) of the wheel. The rest of the water should enter the vanes within the next foot or so, depending upon the size of the wheel. The flume may be constructed of either wood or metal. Aluminum makes a very good metal flume, but wood is preferred. Be sure to install the flume so that it slopes gently downward toward the water wheel. The flume should be bottomless just above the water wheel, as shown in the diagram.

A small wheel, about 5 or 6 feet in diameter, will be sufficient to generate small amounts of electrical power. The wheel rim should be constructed of metal, but wood would serve as a substitute. If you are using metal, be sure to use a non-rusting type, and if you are using wood, use a type of wood that weathers well.

The wheel shown in the diagram on page 120 is constructed with two flat rims welded to a large-diameter pipe and fitted with curved vanes which are welded to both pipe and rims. This assembly is connected to a hub by means of four spokes. The spokes can be made from 1½" or 2" steel pipe welded both to the hub and to the inside of the large-diameter pipe. The hub should be keyed to the drive shaft and tightened with set screws. A "V" groove pulley is fitted to one end of the shaft to drive the alternator.

The alternator can be of the automotive type (12 volt) and must have the proper regulator between it and the batteries. The batteries are connected to a converter, which converts the battery current to 115 volt AC. The diagram shows two 6 volt batteries connected in series, but a heavy-duty 12 volt battery can be used instead.

The shaft and water wheel assembly must be supported by a rugged base made of concrete or wood. The bearings, connected to the shaft, are attached to the base with bolts. These bearings should be good-quality sealed ball bearings.

If the water supply is large enough, a larger water wheel can be used, 10 or more feet in diameter, and can be connected to an AC generator. This type of system will not need batteries, regulator, or converter, but will need some way to control the speed of the water wheel. In order to produce AC current with the proper number of cycles, the speed of the generator must be controlled.

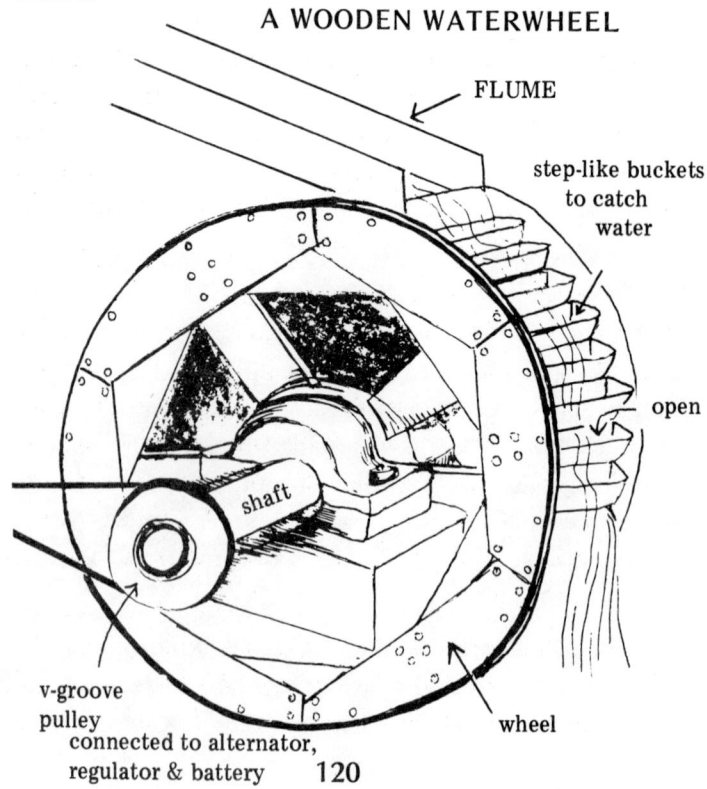

A WOODEN WATERWHEEL

FLUME

step-like buckets to catch water

open

shaft

v-groove pulley connected to alternator, regulator & battery

wheel

WIND

Windmills

Windmills have been in use in Persia, since 915 A.D., when they were used primarily for grinding grain. These windmills were constructed as "horizontal mills", with blades radiating on a horizontal plane from a vertical axis connected to a simple tower-like structure.

The early Romans modified the form of a windmill by developing a "vertical mill", with blades radiating on a vertical plane from a horizontal axis, connected at a right angle to the tower. The earliest type of vertical mill, called a "postmill", consisted of a basic box-like tower which housed the gears and other machinery used to grind grain or irrigate crops. In order to keep the blades facing squarely into the wind, they were turned manually as wind direction changed.

A further development was made with the invention of a fantail, a flat structure attached at a right angle to the blades. This fantail holds the blades directly into the wind, working on the same principle as a weathervane: as the wind hits the side of the fantail, it causes the blade-and-gear assembly to pivot, facing squarely into the wind, and holding it there.

The blades, which may be constructed of wood or metal, are mounted on a shaft. Usually, anywhere from 4 to 20 blades are used and are mounted in such a way as to twist slightly so that they will catch the wind currents. This series of blades is called a "fan".

The fan converts the wind energy into rotating mechanical energy which, through a reduction gearing, operates the crank. The gear box is attached to the end of the shaft behind the fan, and mounted to the stubtower. The crank changes the rotating energy into reciprocating energy, which then drives the pump.

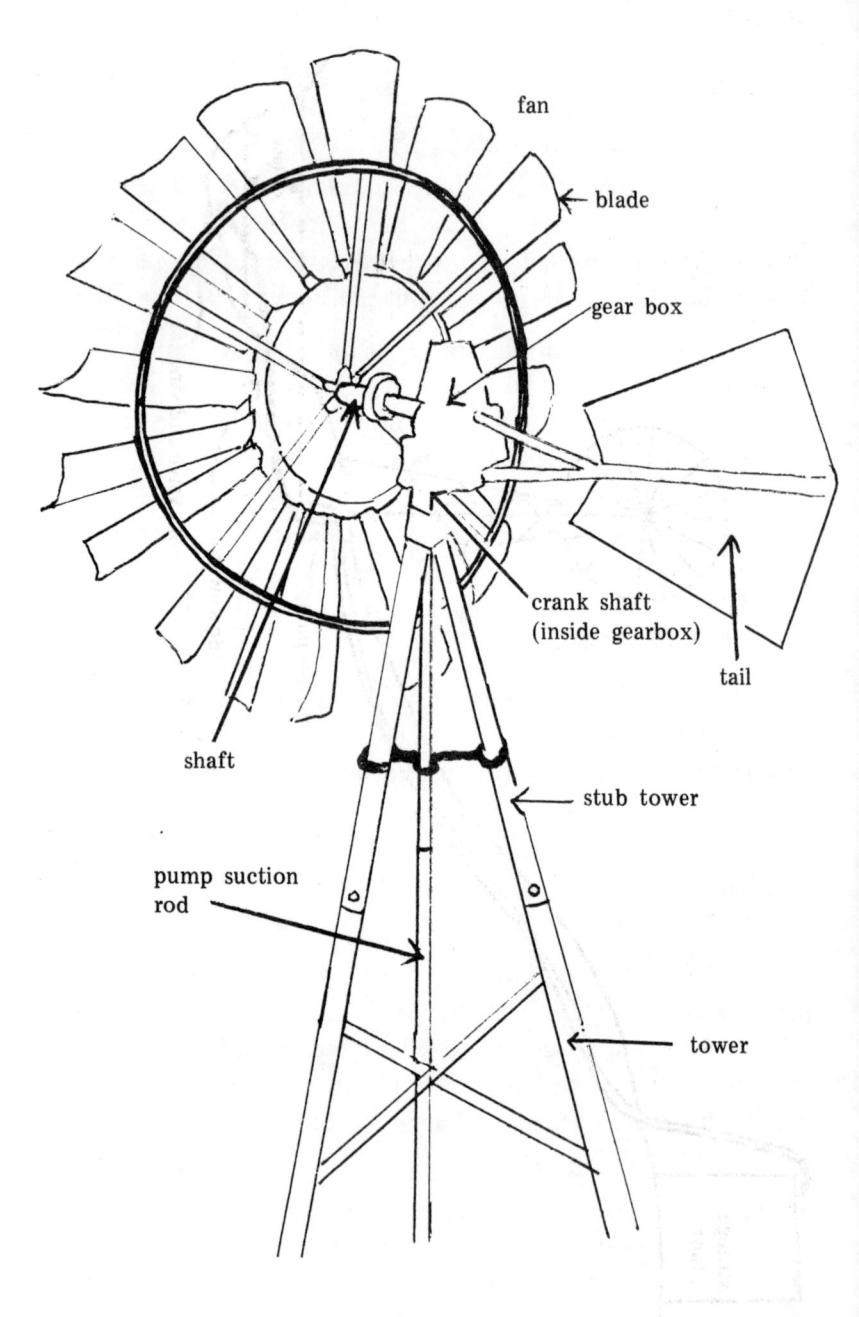

A WINDMILL

122

20' or less

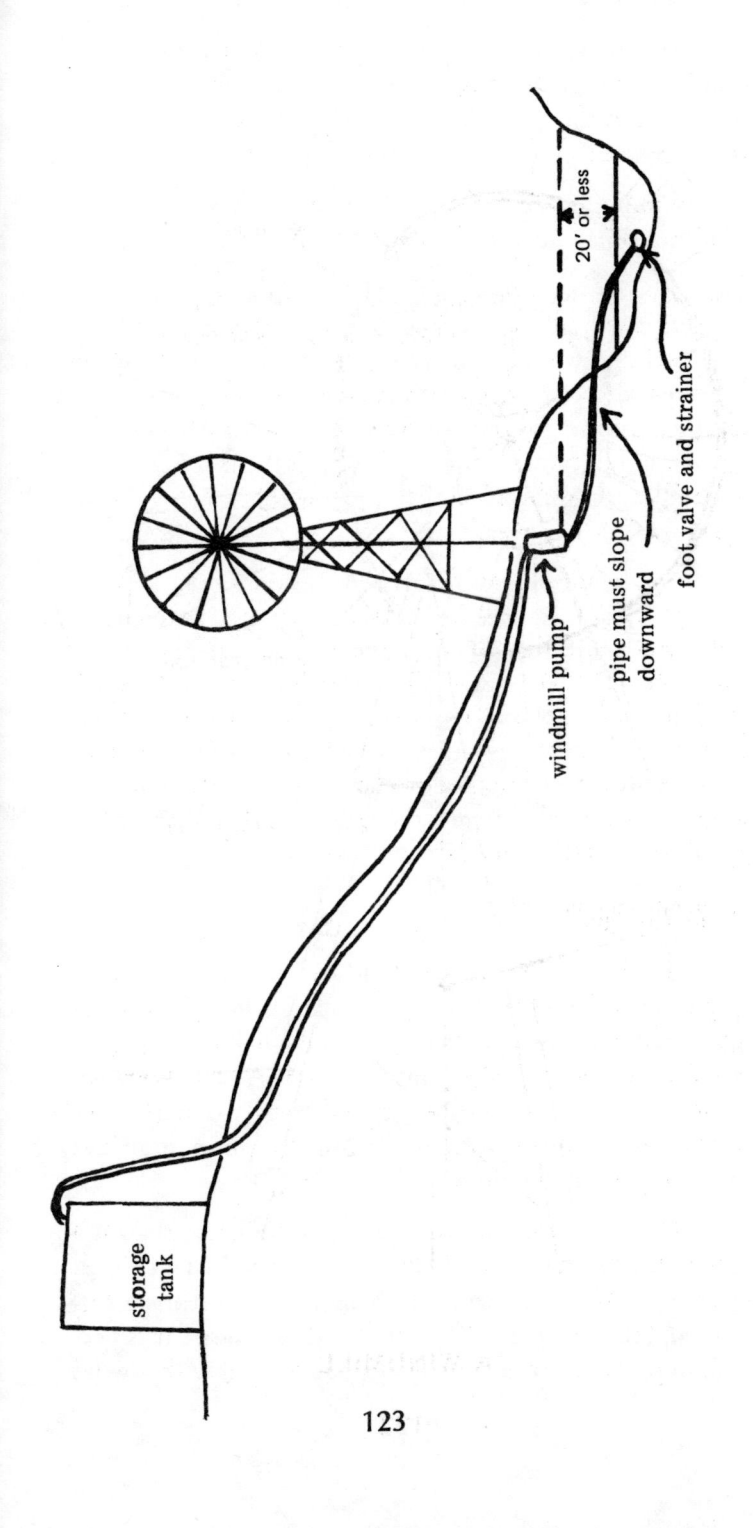

windmill pump

pipe must slope downward

foot valve and strainer

storage tank

The entire assembly is mounted on a tower, generally 20 to 30 feet high. This height is necessary to raise the fan a safe distance above the ground. This tower may be constructed of wood or metal.

Windmills are available in many designs and sizes. Windmills may also be constructed from materials found in junk yards.

The wind-driven water pump consists of a sucker rod, packing gland, cylinder, piston, and a series of valves. The cylinder-piston-valve unit is installed inside the well casing, close to the bottom of the well. Maintenance of the windmill-pump unit generally consists of keeping the mill oiled and replacing the cup leathers on the pump piston when they wear out.

The windmill can be offset from the stream, if necessary, so that it may be located in an area of dependable wind currents. The cylinder must not be located more than 20 feet above the water level of the stream. A trench can be dug from the windmill to the stream and, by placing the cylinder in the bottom of the trench, suction lift can be lessened. Be sure that the suction pipe slopes downward toward the stream and that there are no air traps in the pipe.

Wind Chargers

Wind chargers are wind-driven generators to supply electrical power. The best location in which to set up a wind charger is usually found on high ridges or knolls, canyons, or mountain passes. A sign of heavy wind currents is "wind sculptured trees", that is, trees whose tops bend or lean at the top and whose branches bend or lean in one direction.

The principle and installation is similar to that of a windmill, except that a generator is used in place of a gear box. Wires run from the generator to a storage battery and then to the main switch-and-fuse box. It is necessary to use storage batteries in connection with a wind charger to obtain a steady supply of electrical power. In this way, you will have electricity even during calm periods. In areas where there is continual wind year 'round, these batteries may not be necessary, but it will be necessary to have the wind charger equipped with a govenor to control the speed if you are generating AC current.

Instead of a slow, multibladed windmill-type fan, a 2- or 3-bladed, airplane-type high-speed propeller is used to draw energy from the wind. The 3-bladed format is preferable for efficiency and balance.

A very cheap wind charger can be constructed from a used automobile alternator. These are available at most wrecking yards. A propeller or other means will be necessary to convert wind energy into rotating mechanical energy. The electrical hookup will be the same as the one shown for the small water wheel. For more complete information on wind power read "Wind, Water and Sun" by Orem.

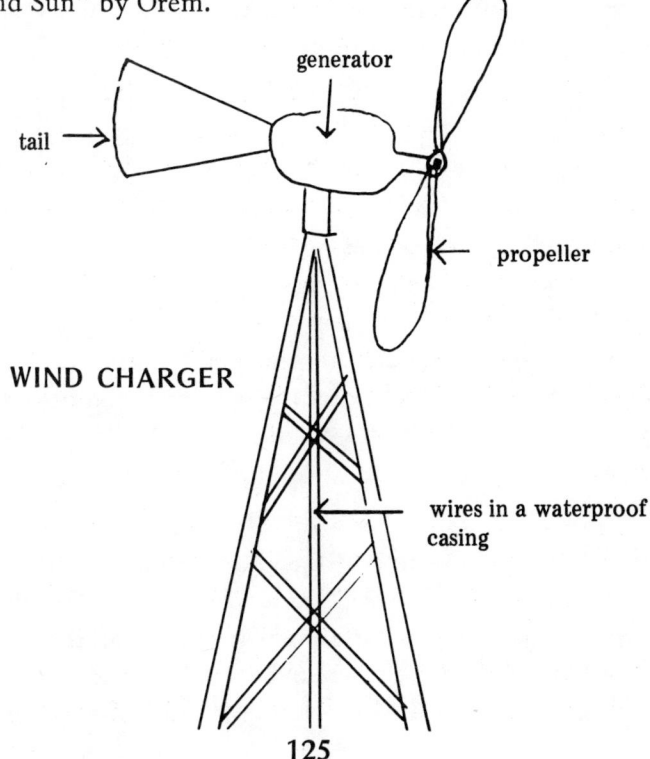

WIND CHARGER

tail
generator
propeller
wires in a waterproof casing

SUN

The sun has always been a source of energy for man. For power, sunshine must be collected, concentrated, and stored. The Incas had discovered a means of collecting the sun's rays to light their sacred fires by using small concave silver mirrors. Convex quartz lenses found in Nineveh ruins were probably used for the same purpose.

In modern days, man has discovered the use of concave (parabolic) mirrors. The sun's rays are concentrated by focusing them on a black holding tank or pipe which absorbs the heat. Water may be heated in this manner, or the water may be converted to steam, which produces mechanical power through the use of a turbine. Parabolic reflectors (which may be obtained by searching an aircraft junkyard for discarded antiaircraft searchlight reflectors) have been known to concentrate the sun's rays to attain a temperature of over 6000° F.

Another way of concentrating solar energy is through the use of a flat "mirror". This collector is composed of several sheets of glass laid over a sheet of metal which has been painted black. An air space is left below each sheet of glass, and a pipe or tubing, filled with water and equipped with an inlet and outlet, is placed between the metal plate and the glass. The metal plate concentrates the solar heat, the tubing collects it, and the glass acts as an insulator and stores it. This may then provide hot water, steam heat for a house, or steam to run an engine.

For heating a house, a solar heat collector may be installed on the "southern slope" of the roof, or as part of a basement "window", as explained later in this chapter. The same method may also incorporate a water heater. Solar heat can only be stored in the form of water in an insulated storage tank for three to five days.

A solar storage battery, or solar cell, converts solar energy directly into electricity. A solar battery is composed of extremely pure silicon, which through a physical reaction when exposed to sunlight, sets up an electric field which produces an electric current. Solar cells live indefinitely, but the cost of refining silicon has made them very expensive to buy. They may be "within the financial reach" of the average person within the next few years.

Possible uses of solar energy are:
    heating a house
    heating water
    generating electricity
    running a commercial furnace
    solar steam engines
    cooking

stones

water heater coil

TOP VIEW

foundation perimeter

SOLAR HEATED HOUSE FOUNDATION

SIDE VIEW

8'

glass

foundation floor
slope ¼" to 1'

cold air ducts
slope ½" to 1'

—S→

Solar Heated House

Solar heating of houses is a very practical and economical way to heat. Solar heat is free and does not use up expendable resources. Solar heat does not pollute the environment.

The method shown here does not have any moving parts and requires no electricity or any other source of power. Rocks under the foundation are heated, using a greenhouse effect. The rocks store the heat while the sun is shining and stay hot for many hours, heating the house from underneath. This beneath-the-floor heat is very comfortable and healthful, and contains all of the heating components in the foundation. This method may also include a means of heating water at the same time.

The foundation will need to be of the basement-type, with a cavity about 4 feet deep. It should have a concrete floor with cold-air ducts built into the floor. The floor and the ducts should slope downward towards the glassed end of the unit. This arrangement allows the warm air to move up the slope of the floor and the cold air to move down the ducts into the area of solar heat. This circulation works on the principle of warm air rising and cold air sinking. The glassed area must be on the south side of the building so that it will catch the low winter sun. There must not be anything to the south of the glassed area that will block out the sun. The ducts are perforated on the top side to allow the cold air to flow down into the ducts.

Soil should be mounded up around the foundation to within about 6 inches of the top and should slope so that water drains away from the foundation.

A black plastic pipe coil can be installed under the glassed area to heat water. The heated water can be stored in a storage tank in the house. The storage tank must have a top connection for connecting the top side of the water heating coil and a bottom connection for connecting the bottom side of the heating coil, as shown in the diagram:

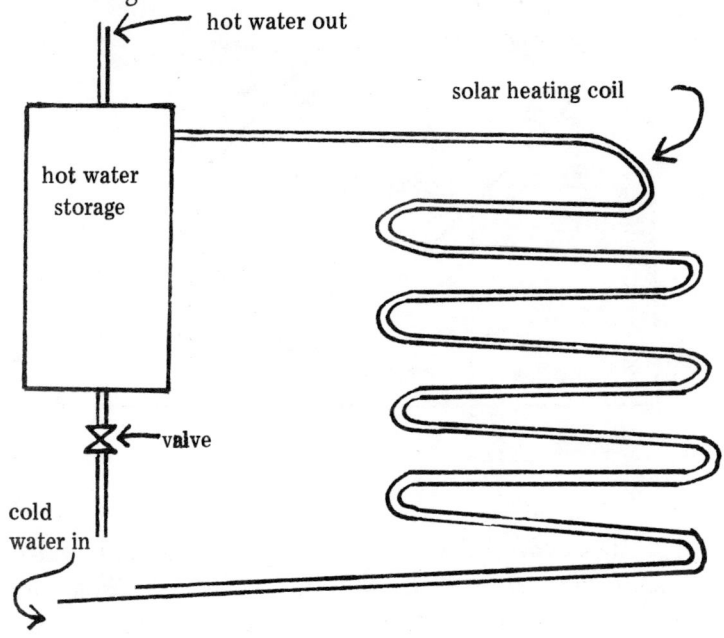

Solar Water Heater

If you live in an area that has a reasonable amount of sunshine, you can have an abundance of hot water with a solar water heater. The type shown here has no moving parts and should operate trouble-free for many years.

For best results, the tank should be made of aluminum, but other materials can be used. The outside shell can be made of ¾ inch exterior plywood, or ferrous cement. The inside of the shell should be insulated for best results.

The tank can be supported by wooden V-blocks about 2 feet apart. These can be made from 2 x 12 lumber about 3 feet long.

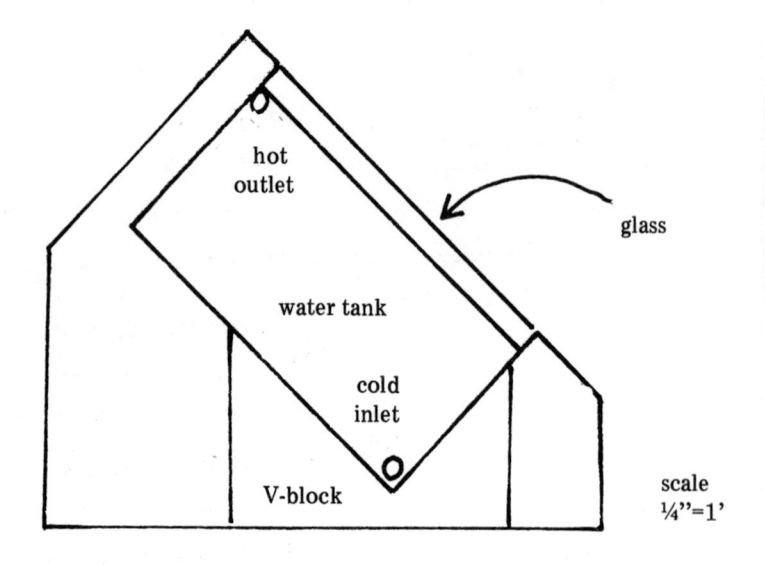

```
        hot
       outlet

                          glass

      water tank

           cold
           inlet

                          scale
    V-block               ¼"=1'
```

The space between the tank and shell can be filled with coarse, clean sand. The sand acts as a heat retainer and will help keep the water hot when the sun is not shining. The top can be fitted with a cover that can be closed when the sun is not shining to help retain heat.

The Woodstove Water Heater

A very simple water heater that may be installed on almost any wood burning stove is shown in the diagram. The heater is operated by closing the bypass damper below the heater. If the water gets too hot, reverse the procedure. Strong support legs will be necessary, and should be welded to the tank. The piping may be fabricated by any experienced pipe welder. A pressure-temperature relief valve may be used as an added safety feature, and is a requirement by most building codes. It allows the escape of pressure from the tank in the event the tank becomes over-heated or the pressure becomes too great.

The Woodstove

If you are installing a wood-burning stove in a house that has combustible floors and walls (such as wood), you should cover them with a non-combustible material in the areas where heat or sparks could cause a fire. It is best not to put the stove any closer than 2 feet to a wooden wall. Where the stovepipe goes through a ceiling or roof, the pipe should be of the jacketed insulation type. The stovepipe should extend about 3 feet above the highest point of the roof. Do not extend the stovepipe out through the roof in a space where there are trees overhanging the house.

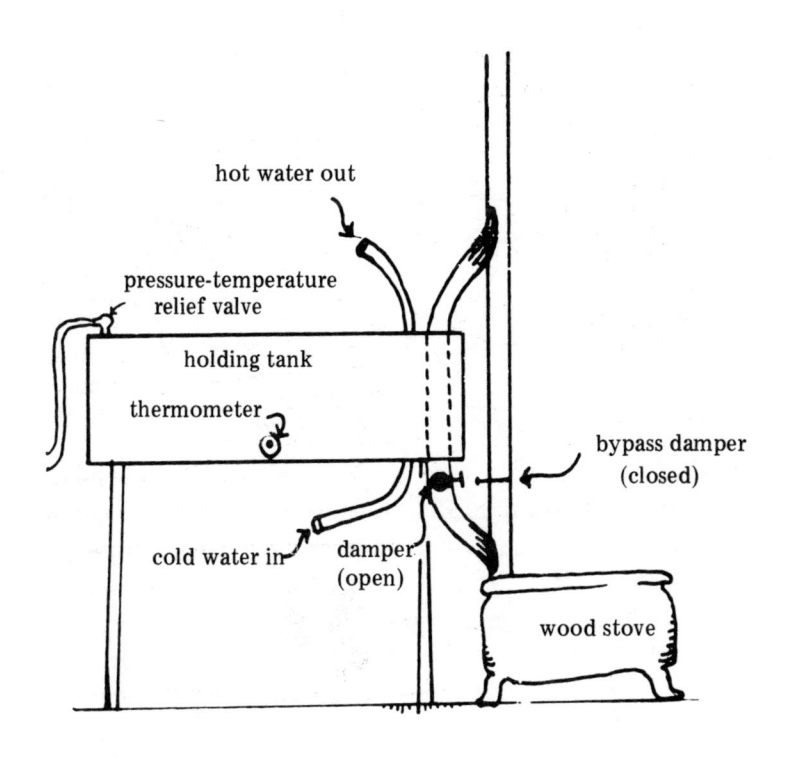

HOW TO START A FIRE

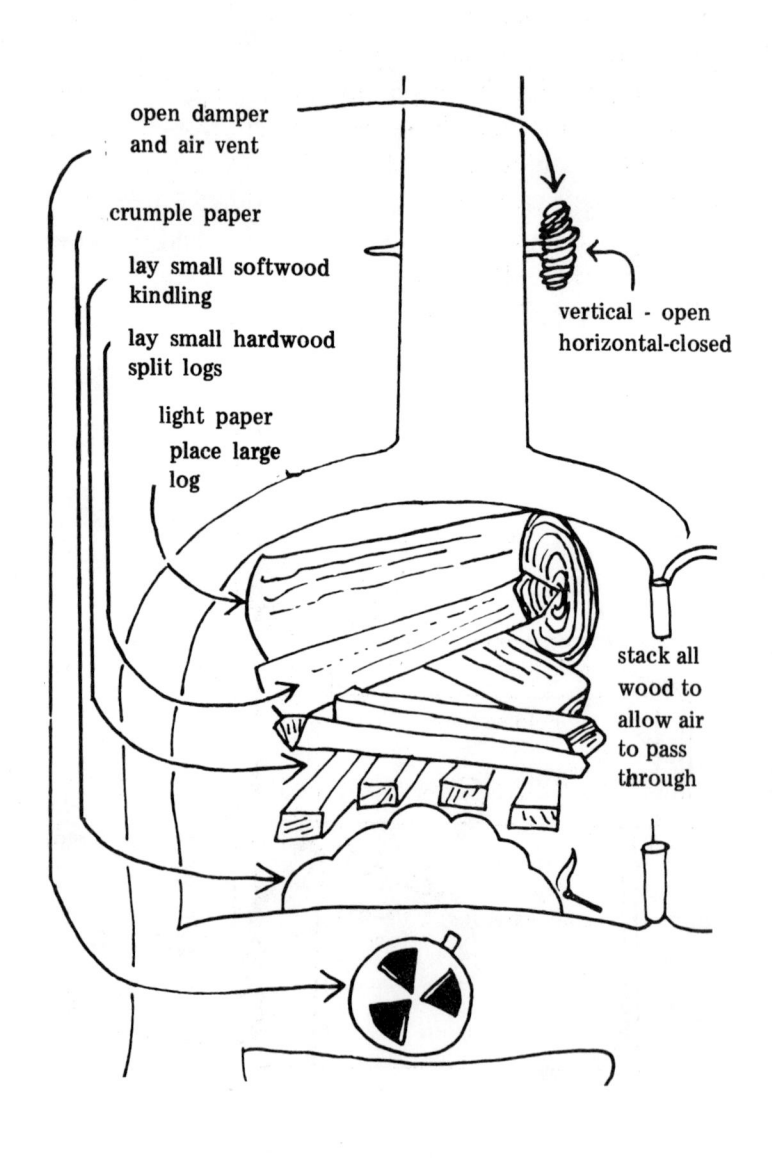

- open damper and air vent
- crumple paper
- lay small softwood kindling
- lay small hardwood split logs
- light paper
- place large log

vertical - open
horizontal - closed

stack all wood to allow air to pass through

PREVENTING EROSION

Erosion, caused by wind, water, construction (such as roads) and other natural and unnatural elements can alter and destroy property, carrying valuable minerals and topsoil away. The following are methods of prevention:

planting trees
planting ground cover (grasses, shrubs, etc.)
damming water spillways to divert water
terracing
improving porosity of soil
planting windbreaks
building windbreaks
limited logging and replanting of two seedlings for each tree cut down
contour farming
ditching
damming erosion gullies
building ponds
installing culverts
mulching

PLANTS TO CULTIVATE WHICH PREVENT EROSION

On Creek or River Bank:

alder	honeysuckle	vine maple
brome grass	spike rush	willow

On Dry, Sandy Creek or River Bank:

beach grass	huckleberry	red cedar
coltsfoot	lupine	sassafras
creeping willow	pignut	scrub oak
gray birch	pitch pine	wild rye
hawthorn		

In Dry Earth:

brake fern   luzula
centipede grass   lyme grass
heather   mouse-ear chickweed
joint fir   sweet fern
legumes

In the Shade:

ferns   ivy   pachysandra
honeysuckle   Japanese spurge   vine maple

In the Open Field:

ash   elm   oak
box elder   hackberry   pine
cottonwood   honey locust   poplar

Windbreak:

arborvitae   gooseberry
Colorado blue spruce   Norway spruce
Douglas fir

General All-Purpose:

alder tree   fescue   pachysandra
alfalfa   heather   pearlwort
Allegheny spurge   honeysuckle   peas
beans   Kenilworth ivy   rape
beggarweed   lentil   rye
buckwheat   lily turf   soybeans
clover   locust   vetch
creeping nettle   millet   Virginia creeper

STOPPING STREAMBANK EROSION

    Whenever water moves rapidly through a channel, it is very likely to carry with it the earth that forms the

STOPPING EROSION OF A STREAMBANK WHERE RIVER OR CREEK IS UNDERCUTTING BANK

stream flow

rock dike

silt deposit

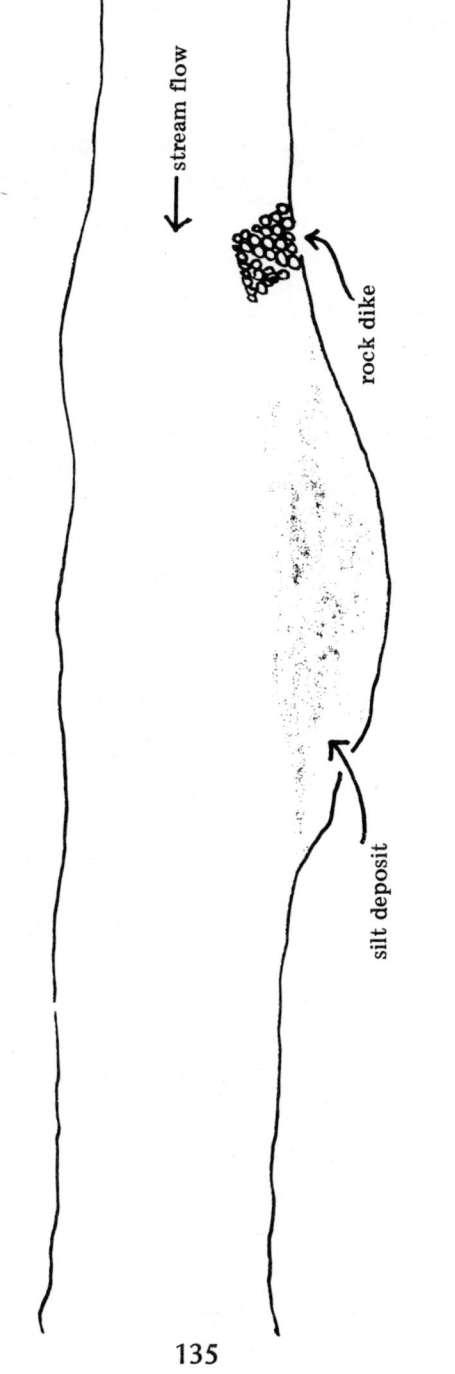

TOP VIEW

banks of the creek or river. The following method is simple and works very well.

Build a rock dam on the upstream end of an undercut bank. The rocks should be fairly large so the water current does not wash them away. They should be piled high enough that a rise of water in the streambed will not go over the top of them, and should extend out into the stream as far as it is feasible.

The water will go around the dam, leaving a still area on the downstream side of the dam. Sand and silt will settle out in this area and will build up a new bank.

If the undercut is very long, it may be necessary to build more than one dam. This method works under the principle of letting the stream do the main part of the work for you. After the undercut is filled with sand and silt it should be planted with water-loving trees, such as vine maple, alder, or willow.

DAMMING EROSION GULLIES TO MAKE PONDS

Some use should be made of the waste land in erosion gullies. One such use is ponds. This will require building dams at intervals of about one for every 10-foot drop in elevation of the land. Concrete or rock dams are the most satisfactory type of dam for this kind of pond. This type of pond may, in a few years, become

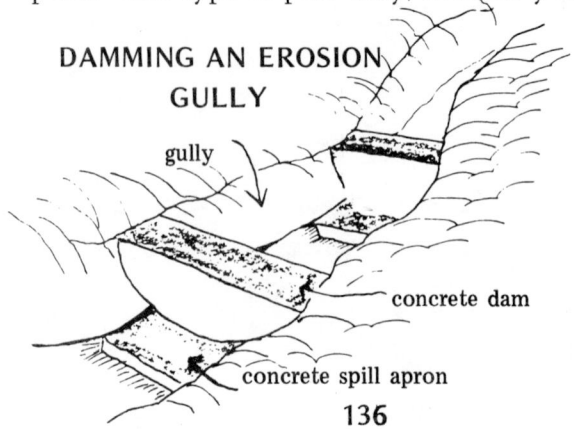

DAMMING AN EROSION GULLY

gully

concrete dam

concrete spill apron

filled with silt. When this happens, the silt may be removed during the dry season. This silt is a very valuable asset to any gardener or farmer. This type of silt is generally very rich in soil nutrients and the texture makes it a very good growing medium. An alternative is to leave the pond full of silt and plant it to some kind of crop. Either way would make beneficial use of wasteland and would slow erosion, sometimes halting it altogether.

HOW TO PREVENT A FIRE

Fire may be started by:
1. Glass bottle, especially when filled with water, sitting in dry grass in the sun—it acts as a magnifying glass.
2. Idling car in dry grass meadow—the exhaust is very hot and may start a fire.
3. Cutting wood with a chainsaw during hot, dry months—exhaust is very hot.
4. Oily rags, combustible liquids stored in glass jars, or paper-wood-cloth, etc., stored too near a stove or fireplace.
5. Paper or rags near a tin roof in the attic.
6. Bad electrical wiring.
7. Unconscious disposal of cigarettes, careless trash and underbrush burning.
8. Improper installation of woodstove pipes.
9. Carelessness.

How to be Prepared:
1. Cut a firebreak around dwellings.
2. Have water reservoirs, such as ponds, water tanks, or 55 gallon drums around farmstead as an emergency water supply.
3. Clear dry branches from roof and around house.
4. Be careful!

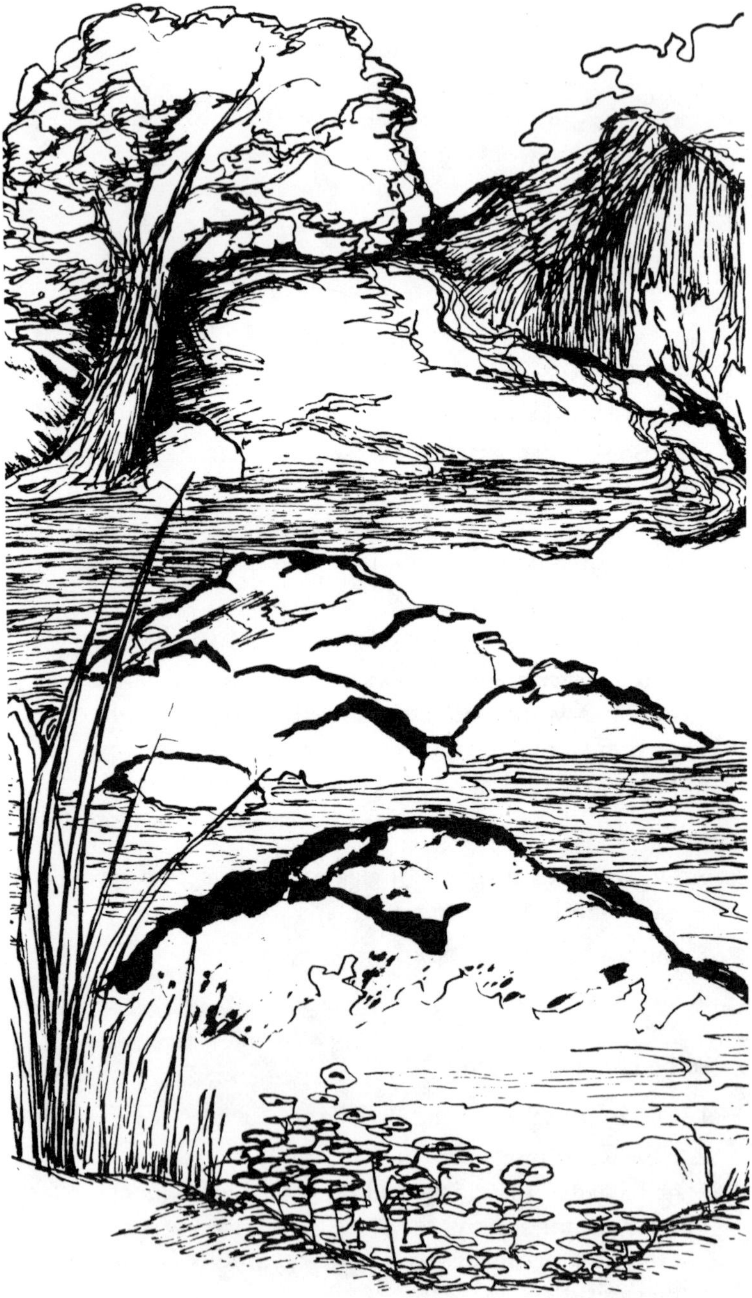

CHAPTER III
DEVELOPMENT OF WATER

THE SOURCE

Water covers vast areas of the earth's surface. It may take the form of ocean, sea, lake, pond, river, creek, spring, or underground reservoir.

A watershed is an area of terrain that collects precipitation, such as rain and snow. This water flows downward toward a valley and some of the water soaks into the ground, entering the permeable stratums of topsoil, sand, and gravel. The water seeps down through these stratums until it reaches an impermeable stratum, such as clay or bedrock. At this point it is halted and is stored as groundwater.

This groundwater collects and continues to flow downward beneath the earth's surface until it reaches its lowest point, such as a valley floor, or until it reaches a vertical impermeable underground structure, such as a hill, an area of clay, a fault, a dike, or bedrock. This structure forms a natural barrier which blocks the water from continuing its downward course. Instead, it is collected in the form of an underground reservoir.

Water may migrate many miles or just a short distance before being confined by one of these barriers.

140

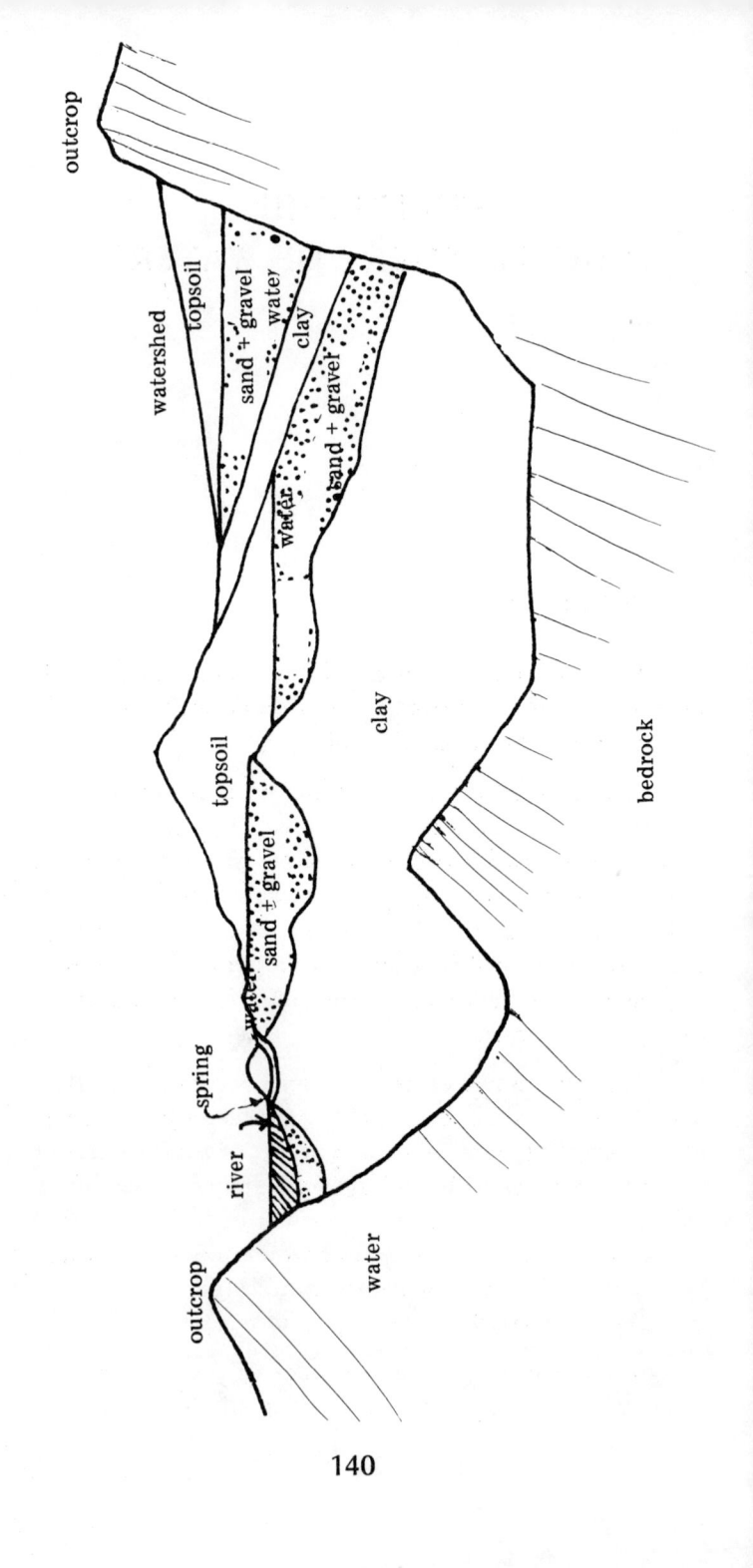

The size of the reservoir will be influenced by the surface terrain, amount of precipitation, and the arrangement of porous and nonporous stratums. When the conditions are such that the groundwater reaches the surface, a spring is formed. Springs continue their downward flow toward the valley, feeding creeks, which feed rivers, which feed oceans, which feed springs with rain. And so flows life's water cycle.

WATER PRESSURE
TABLE W-1

The water pressure at any height may be found by multiplying the height in feet by .43. This will be the pressure in pounds per square inch at the house.

Ht. of Tank above house (ft.)	Pressure at house (lbs.)	Pipe size (in.)
10	4	2
20	9	2
30	13	2
40	17	2
50	22	1½
60	26	1½
70	30	1½
80	35	1½
90	39	1

CONSTRUCTING A WATER TANK

Select a building site for the tank which is at the right elevation for the desired water pressure (see Table W-1). The tank should be about 90 feet above the house to give 39 pounds of pressure at the house. If you do not have a suitable location this high above the house, a lower tank site will work satisfactorily by increasing the size of the supply pipe from the tank to the house. If the tank is 25 feet above the house, you will need a 2 inch supply pipe. If it is 50 feet, you will need a 1½ inch pipe. If it is 90 feet or more, you will need a 1 inch pipe. To increase quantity of water flowing through the pipe, use a section of pipe at the springhead or tank which is one size larger than recommended in Table W-1. About half way down the line, reduce pipe size to the one recommended, creating a funnel effect.

Table W-1 may also be used to determine the pressure and required pipe size when hooking up a water system directly to a spring, creek, lake, or any other body of water.

If the tank site is not level, it will be necessary to level it. The site must be of a solid, stable material, preferably bedrock, rocky soil, or packed sand, though hard-packed clay will suffice.

CAPACITY OF TANKS
TABLE W-2

A very simple way to estimate the capacity of a tank is to multiply the diameter x diameter x height x 6. This will give the approximate capacity in gallons of the tank.

Tank Diameter (ft.)	Tank Height (ft.)	Tank Capacity (gal.)
5	1	150
5	2	300
5	4	600
10	1	600
10	2	1200
10	4	2400
15	1	1350
15	2	2700
15	4	5400
20	1	2400
20	2	4800
20	4	9600

TOP VIEW

center stake

½" reinforce-
ment bar
on center-
line circle

SIDE VIEW

½" or ¾" re-
inforcement
bars

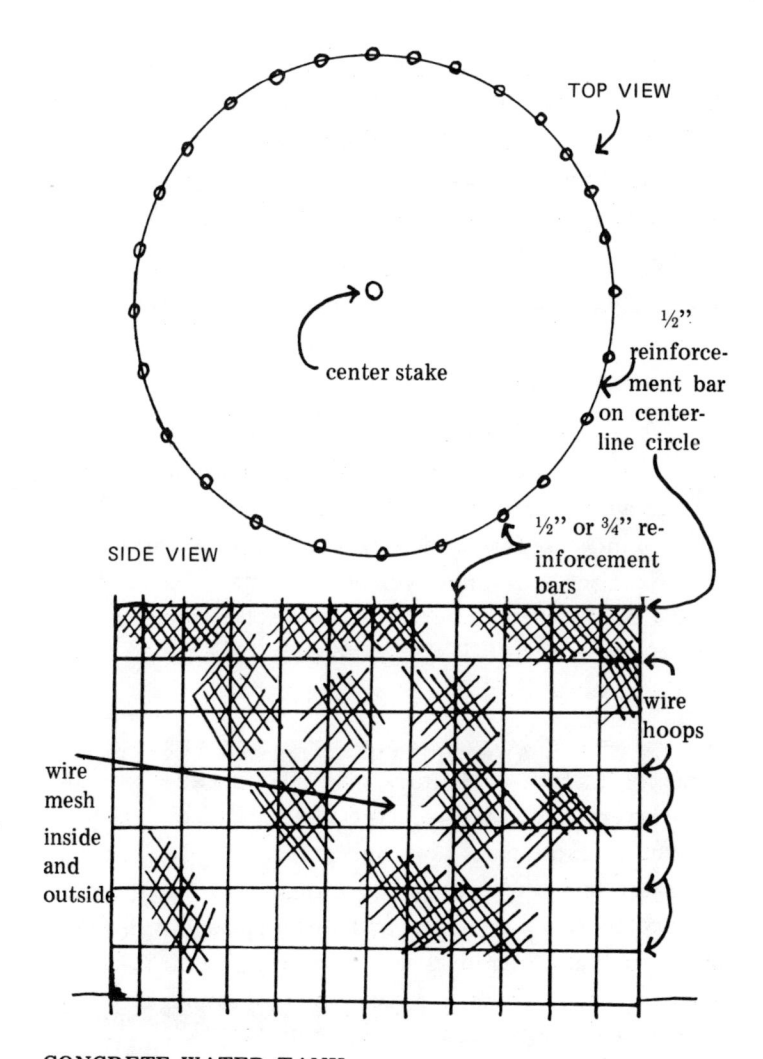

wire
hoops

wire
mesh
inside
and
outside

CONCRETE WATER TANK

Select required size of tank from Table W-2. Drive a stake (a round steel rod is best) where the center of the tank will be. Tie a loop in one end of a strong cord or wire. Place loop over the stake. Measure along the cord a distance equal to the radius (half the diameter) of the tank. Tie a sharp stake at this point on the cord. Using the point of the stake, scribe the circumference of the tank. Be sure to hold the stake vertical.

Cut reinforcement bars about 1 foot longer than the tank height. Drive the bars vertically around the scribed circumference line about every 2 feet, leaving them about 1 inch shorter than the finished tank will be. Reinforced concrete is also called ferrocement.

Bend a ½-inch reinforcement rod to fit the centerline circle of the vertical bars. Fasten the circle rod to the vertical rods by tying in place or by welding. Welding is definitely the stronger of the two.

Wrap ½-inch-mesh wire around the outside of the vertical rods twice, forming a double layer on the outside, and fasten with tie wires at each vertical pole. Then wrap ½-inch mesh wire around the inside of the vertical rods, forming a double layer on the inside. Fasten with tie wires. Bind the mesh horizontally (around the outside of the circle) with 10-gauge galvanized wire every 6 inches (vertically). These will act as "hoops".

After the 10-gauge wire hoops are in place, the concrete bottom of the tank may be poured. If you are using river sand and gravel, the mix should be one part cement to three parts sand and gravel. If you are using separate sand and gravel, the mix should be one part cement to two parts sand and three parts gravel. Add water so that concrete is just wet enough that it can be worked easily. Pour a 3-inch layer of concrete in the bottom and smooth it with a 2 x 4 or a trowel.

After the concrete in the bottom has set overnight, the sidewalls may be plastered. The plaster is made by mixing one part cement to two parts sand and adding just enough water to make it workable. It should be as dry as possible, because if it is too wet it will fall off of the mesh wire. This job will require at least two persons doing the plastering and one person to mix the plaster. Two people work at the same spot, facing each other,

weld

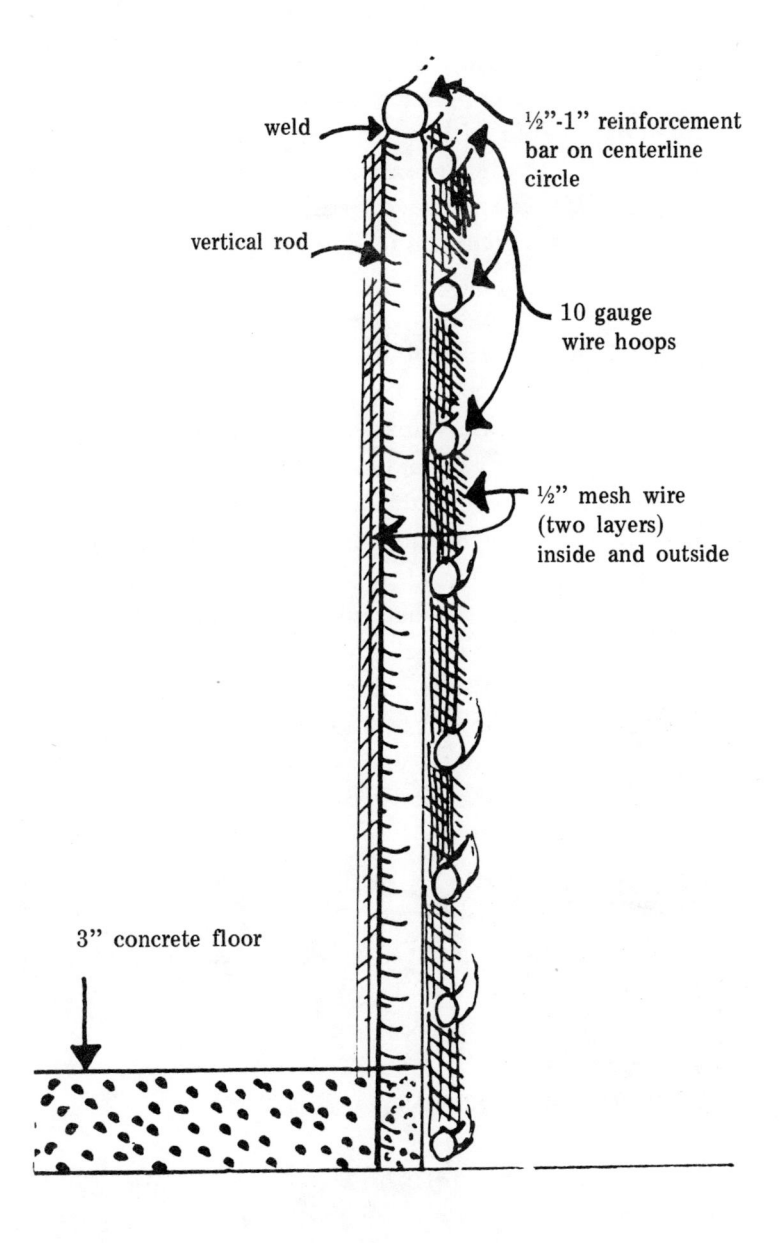

vertical rod

½"-1" reinforcement bar on centerline circle

10 gauge wire hoops

½" mesh wire (two layers) inside and outside

3" concrete floor

one working inside the tank, and one working outside the tank. The plastering is done with a trowel, working it into the mesh thoroughly so that a solid wall is formed.

When finished, the tank wall should be about 2 inches thick. Keep plaster moist for about a week by spraying it often with water. Finish sealing the tank by painting the inside with a mixture of cement and water mixed to the consistency of stiff paint. When the tank is thoroughly sealed, it is ready to be filled.

WOODEN WATER TANK

Redwood water tanks may be purchased new and may also be found second-hand. Wooden tanks may be constructed from "scratch" of redwood 2 x 6 lumber. Wooden tanks are not recommended, because they are subject to shrinkage unless they are kept full of water most of the time. If the tank remains partially empty for any length of time, the wood will shrink above the water line, causing tremendous leakage. It will be very difficult to stop the leaks. A tight-fitting cover will help prevent the tank from drying out by keeping the humidity high inside the tank. Never seal a tank with tar or other toxic products.

OTHER WATER TANKS

pond reservoir	concrete pond
stone and mortar	used wine barrels or vats

DEVELOPING A SPRING

A spring is an outlet for water that is confined in a reservoir of some kind. Usually the reservoir is underground. It may be sand, gravel, or any other porous material, or it may be a cavity in limestone or other rock. Water is generally trapped in the reservoir during rainy periods and is released through the spring over a period of time. The capacity of the spring depends on porosity of the rock and surrounding earth, climate, amount of rainfall, and capacity of the reservoir.

Sometimes the flow of a spring can be increased by enlarging the discharge opening of the spring. Many times a large part of the spring water is lost because it sinks into a gravel or sand bed before reaching the **discharge** opening of the spring. By properly developing the spring, most of the water can be collected.

The first step in developing the spring is to clean out the opening of the spring. Then clean out several feet in all directions from the opening, if possible. This will give you a chance to observe the free flow of the spring.

Select a length of pipe large enough to carry the full flow of the spring. This will be used temporarily to keep the area dry while you are working with it. Place one end of the pipe in the spring opening and seal around the pipe with clay. It may be necessary to build some small clay dams across the pipe to channel all of the water through the pipe.

After you have all of the water flowing through the temporary pipe, you are ready to begin the trench for the spring box. A semicircle shape is generally best for recovering all of the water. Each spring is different, so it is not possible to give a complete outline of the spring development. By using these instructions as a guideline, you should be able to tap the spring successfully.

The box trench is 4 to 6 inches wide in a semicircular shape, and dug deep enough to reach an impermeable material such as rock or clay, if possible. The trench should be dug far enough away from the spring opening so as not to restrict the flow of water from the spring.

After the trench for the spring box is completed, dig a trench for the discharge pipe perpendicular to the semicircular trench. The pipe trench should be dug to the depth of 1 foot above the bottom of the semicircular trench and should slope downward away from the spring box. Place the discharge pipe in the trench so that one end enters the semicircular enclosure and the other end slopes downward away from the spring box and into a water reservoir or tank. The end of the discharge pipe nearest the spring should be plugged with something to prevent the pipe from filling with mud or concrete during the construction of the box.

Refill the pipe trench with earth, covering the discharge pipe and packing it solidly, so that the semicircular trench may be filled with concrete. Mix three parts sand and gravel to one part cement with just enough water to make it workable. Pour this concrete into the semicircular excavation.

After the concrete has set overnight, excavate the semicircle and pour a 2-inch layer of concrete in the bottom. Take care not to cover the discharge pipe with concrete. When the concrete in the bottom has set overnight, remove the temporary pipe from the spring opening and the material used to plug the discharge pipe.

Pack more concrete on top of the semicircle edge of the spring box. This is usually a mound 6 inches above the level of the earth surrounding spring box to prevent the earth from caving in. It also provides a place for the spring box cover to be set on.

A cover for the spring box can be built by covering a flat surface with heavy paper and placing a suitable

TOP VIEWS

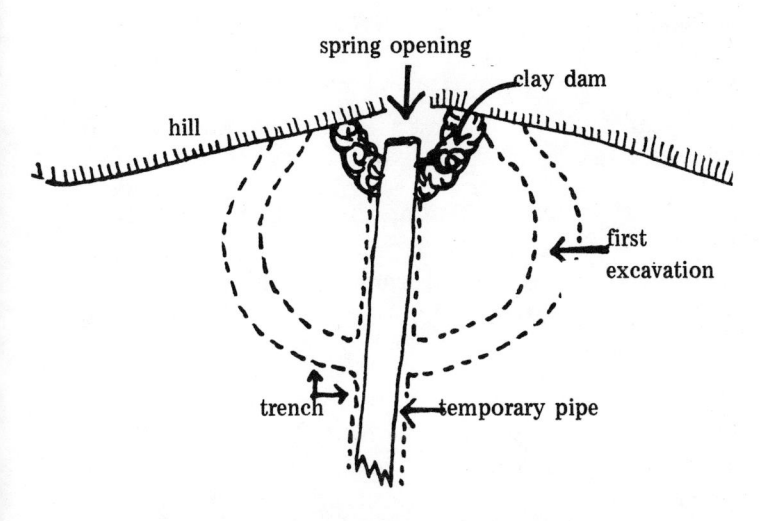

spring opening
clay dam
hill
first excavation
trench
temporary pipe

Ready for First Concrete Pour

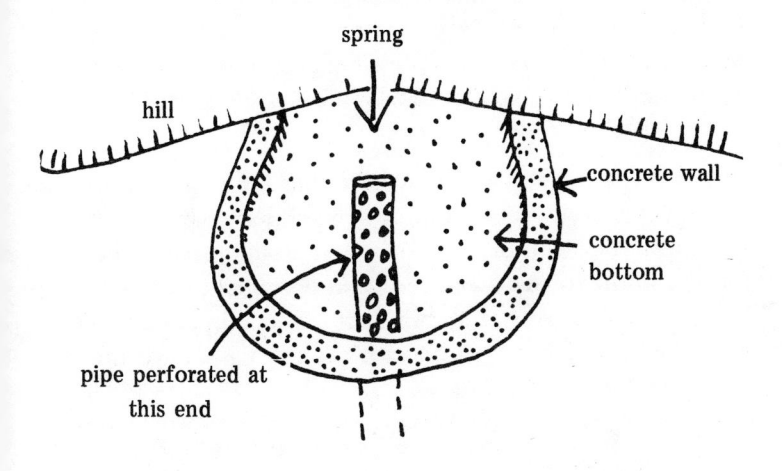

spring
hill
concrete wall
concrete bottom
pipe perforated at this end

Completed Spring Box

149

form over the paper. The form can be made from plywood or flat bar. Pour concrete into the form. Pour it thinly enough that the cover will not be too heavy. Let concrete cover set about one week, keeping it damp, before placing it on top of the spring box. After the cover is in place, a 4 or 5 inch layer of earth may be put over it to seal it, or you may seal around the outside of the lid with a thin layer of plaster.

Because concrete is nonporous, it is the best material to use for a spring box, as it catches and holds a large percentage of the water flowing from the spring. This is especially important when tapping small springs and during the dry season.

The shape of the spring box may vary according to the area you are working with. Other practical shapes are square, V-shaped, or any shape that will confine the water.

The spring box may also be built out of a variety of materials, such as wood, stone, brick, clay, or metal. It should be built as tightly as possible, sealed if necessary, nonrusting and nonpolluting, and should contain a bottom. When using wood, use one that does not decay rapidly, such as redwood or cedar.

Except in the case of springs that contain an abundant water flow year 'round, the discharge pipe from the spring box should lead to a large holding tank. This may be in the form of a water tank, pond, or other reservoir. If your spring has abundant water all year 'round and your water needs are small, the discharge pipe may lead directly to the house.

DEVELOPING A CREEK

Generally, a creek is not a good source of drinking water because it is open to all kinds of pollution. If you own the whole watershed of the creek, you may be able to keep it clean enough for drinking purposes. Do not allow animals to graze upstream from your water supply. If the creek does not contain any chemical pollutants, it is generally good for most uses other than drinking water. It is a good idea to construct a road to the stream in the area where the water will be developed. This will allow materials to be easily moved to the site and will make maintenance of the system much easier.

Extracting water from a creek is a simple matter when the creek is above the level of the area to be fed by the water. A pipe is run from the creek directly to the house or to a holding tank above the house. The end of the pipe which goes to the creek should be mounted so that it is stable during seasons of rapid water flow. This end should also have a strainer to filter out dirt and leaves which might clog the pipe.

Many of the problems with a creek water system are caused by the strainer on the intake pipe, as it may easily become clogged with earth sediment. The strainer should be installed parallel with the creek and in an area of rapid water flow. This rapid flow of water over the screen surface will keep the strainer free of leaves and other debris which are the main cause of failure in this type of system. If there is not an area of rapid flow in the stream, one can be created by partially damming the stream with rocks.

If the water is deep enough, it is also a good idea to install and mount the strainer so that it is set a few inches above the creek bottom. This will prevent any sediment from clogging the strainer. The strainer should be anchored solidly to prevent it from being washed away. The pipe should slope downward away from the strainer and should continue the downward slope to the point of discharge.

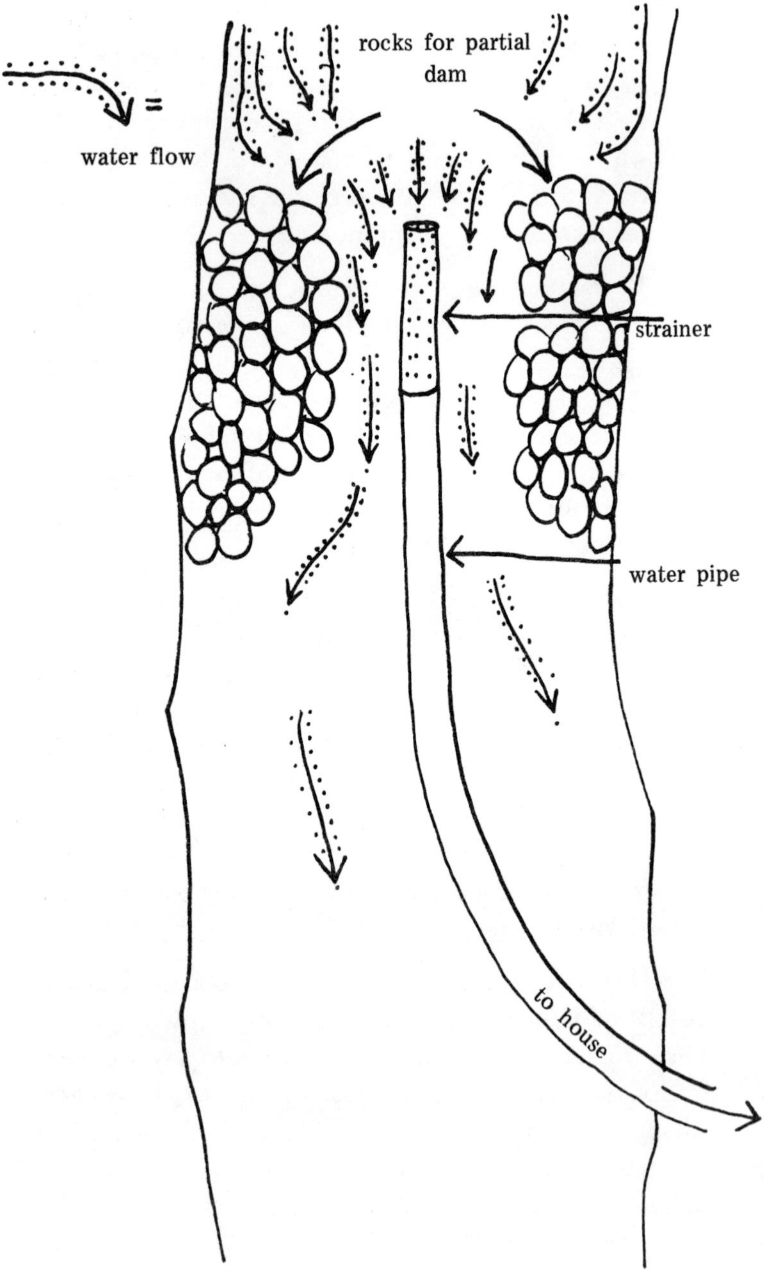
rocks for partial dam
= water flow
strainer
water pipe
to house

Never use any kind of detergent or soap—not even biodegradable soap—in creeks, ponds, or other waters. It will cause great damage to fish, birds, and other wildlife.

Creeks may be dammed to provide a larger water supply or to form swimming holes. This will be explained in the section on Pumps of this chapter.

Creeks may provide power for generating electricity. This is explained in Chapter II.

A CREEK WELL

A creek well should be located to one side of the stream bed. This is to prevent debris from washing into the well. The well should be deep enough to be well below the lowest water level in the dry season.

The creek well liner should be of concrete. This liner is a cylinder, perforated throughout the bottom half. This perforated half goes below the low water level.

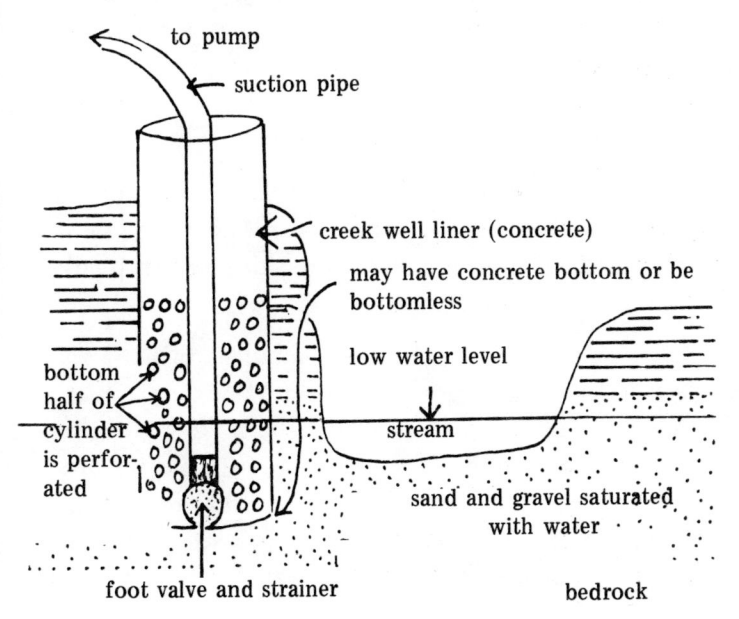

PONDS

When building a pond, choose the location carefully. The more level the earth, the smaller the amount of earth that will have to be moved. Some terrain lends itself naturally to pond construction. In areas that have a high slippery-earth composition (such as blue clay), the pond should be planned very carefully so that erosion does not fill in the pond and cause hillsides to slide and cave in. If the pond is to be used as a source of irrigation water, it is best to have it above the site to be irrigated to provide for gravity-flow of water. If placed below the site to be irrigated, it will have to be pumped up to the site. A cool spot, such as under trees or on the north slope is best for a fish pond, whereas a swimming hole would be more pleasant in a sunny place.

It is best not to have a creek flowing directly into the pond, as sediment brought by rapidly moving water will gradually fill up the pond space. The ideal situation is to have the pond fed by one or two springs. This allows circulation of the water and prevents the water level from getting too low in the summer. Springs do not move as rapidly nor in as great volume as creeks, and will add very little sediment to the pond.

It is also possible to build a series of ponds at various levels, the highest pond fed by a spring, the second highest pond fed by the highest, the third highest pond fed by the second highest, and so on.

Size and shape of the pond should be planned to fit your needs and the surrounding natural terrain. Once the location has been chosen and the pond has been planned, the next step will be to remove the earth from the spot. Depending on job size, a bulldozer or other earth-moving type of machinery may be used or it may be done by hand with a pick and shovel.

In digging, you may hit an underground spring or water table, which will be of great advantage in keeping the water circulating. The water plants mentioned in Chapter I may be an indication of this underground water source. If you do not hit an underground source of water, you will have to rely on the spring and the rainy season to fill the pond. It is best to dig a pond in the fall just after the first rains begin. This will help to pack the earth inside the pond to seal it. A 6-inch layer of pure clay will be sufficient to seal a pond. If your area is dry and/or the soil is of porous material, such as sand, you will need to seal the pond so that the outflow of water does not exceed the inflow. This is done by packing a 6-inch layer of pure clay or other nonporous earth material around the inside surface of the pond. If you are in an area where the earth composition is fairly nonporous, you may seal the inside pond surface by smoothing the inside with water. As the pond fills, natural sediment will seal the minute cracks.

PONDS WITH NO FLOW-THROUGH WATER

This is possible, contrary to popular belief, and can be the saving grace on property with little or difficult-to-develop water systems. It opens a wide range of possible swimming holes, reservoirs, and fish ponds. A pond near a garden can actually alter climate and lengthen growing season, through the evaporation of moisture and heat retention. It provides homes for various beneficial animals and attracts birds. A pond without flow-through water may be constructed nearly anywhere on the property.

The development of this type of pond is based upon setting up a harmonious balance with nature, similar to setting up a balanced aquarium. By conscientiously introducing the proper plants and animals, an environment is created that can take care of itself through constant purification and regeneration.

Algae, erroneously thought by many to mean pollution, are the base of the food chain. In manufacturing their food, algae release oxygen, adding oxygen to both the air and the surrounding body of water. Without algae we would have no oxygen to breathe, as 90 percent of the world's oxygen comes from algae in the ocean. Algae also provide food for various pond animals.

Algae in ponds, which form a greenish spongy growth along the edge are actually a sign of a healthy pond. It may spread rapidly, especially when the surface of the pond is in full sunlight. If it begins to "take over", it may be removed and added to the compost heap where it will add many minerals and humus.

A pond should be built so that it has both shallow and deep areas so that the various plants and animals that require various water depths will be able to survive. The deep part of the pond should be from 10 to 20 feet deep, and the shallow parts should be about 3 feet deep, tapering to a thin edge at the shoreline. This shallow area should extend around the perimeter of the pond and be at least 10 to 20 feet wide to allow plenty of room for shallow-water plants and animals.

It will be necessary to seal the pond from leaks (1) in heavy clay soil by packing and smoothing the surface inside with water, which seals the cracks in a way similar to plastering, or (2) in porous soils by spreading a 6-inch layer of pure clay over the inside surface. Soil porosity may be tested, using the soil test mentioned in Chapter I.

When the pond is to be used for irrigation or other utility water, it should be built deep enough to hold all of the water which falls in a year, or deeper if possible.

If, for instance, the average rainfall is 72 inches per year, the pond should be at least 6 feet deep, though the deeper the better. It is excellent to dig down to bedrock, as this bedrock acts as a natural sealer on the bottom. The pond should be dug so that the sides slope gently inward toward the center at the bottom.

PONDS SEALED WITH CONCRETE

The method of sealing a pond with concrete is especially useful in areas with a very porous soil where the import of clay would not be practical. It also works well for small ponds and ponds being built to act as irrigation water storage reservoirs.

First, the hole is excavated. Next, place 1-inch-mesh chicken wire around the cavity surface so that it conforms to the natural contours of the pond. Make a mixture of one part cement and two parts sand with enough water to make it workable. Beginning at the center in the bottom, spread a 1 inch layer of concrete and work your way up the sides of the pond. Keep this layer damp by spraying with water for about three days. On the third day, spread a ½-inch layer of the above concrete mixture over the first layer, working your way up from the bottom. Let set for one or two days, then paint the inside surface of the pond with a mixture of cement and water mixed to the consistency of stiff paint.

When this sealing layer is dry, cover the inside area of the pond with a 2-inch layer of earth. This will provide rooting for plants, which are essential to the balance of the pond.

PONDS SEALED WITH BENTONITE

It is best to apply bentonite during construction when the soil is dry. After the area to be sealed is leveled the bentonite is spread evenly over the surface about ½ inch thick and covered with 4 to 6 inches of topsoil. When the bentonite becomes wet it expands, making a very tight seal.

For balanced ecology in a pond with or without flow-through water, a variety of plants and animals may be introduced into the pond:

Surface Plants:

azolla   duckweed

Shallow Water and Along Submerged Bank:

arrow arum   flowering rush   primrose willow
blue flag   golden club   spike rush
cattails   lizard tail   umbrella palm
coontail   marsh marigold   water poppy
elephant's ear   pickerel weed   wild rice

Plants which add Oxygen to Water (plant in shallows):

arrow head (sagittaria)   elodea
cabomba   tapegrass (vallisneria)
coontail   myriophyllum

Plant in Deep Waters (3-6 feet):

American lotus   cowlily
Cape pondweed   water lettuce
coontail (hornwort)   water lily

Animals:

catfish (do well in warm, murky water as they are very hardy. Do not need much oxygen. Also enjoy clear fresh water. They are scavengers.)
crayfish
fresh water clams.
gambusia (top minnow or mosquito fish) eat mosquitoes
pond snails keep water clear

trout (should be introduced only into deep, cold water.) water flea (zooplankton or daphnia) keep water clear.

The above plants and animals may be available locally in the wilds, or may be purchased through fish hatcheries, wildlife nurseries, or large aquarium dealers.

RAINWATER

Rainwater is the purest form of natural water available. It is naturally distilled and is healthful and useful for many purposes.

It may be collected during the rainy season in barrels or drums. An open reservoir, such as a pond or water tank, may also be used to collect rainwater.

A simple method of collection is to build rain gutters on the roof of a house or barn which angle downward at a slight slope toward a storage tank. Gutter spout should be installed so that the rainwater may be diverted outside of the storage tank when necessary. At the onset of the rainy season, allow rain water to flow outside of the storage tank during the first few rains to wash any dust and dirt off of the roof.

This method should be used only on roofs that are built of wood, or other nonpolluting materials.

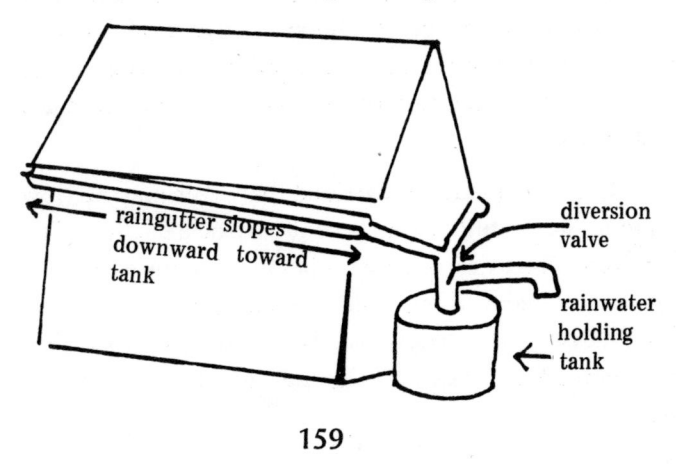

DEVELOPING A WATERSHED

There appears to be a great shortage of pure water in many parts of the world, and yet in many places there is sufficient rainfall to furnish all the needs of an area if there were proper storage facilities. Ground storage is the most economical. If you are in a hilly area, you may be able to store the water on a hilltop and have gravity-flow pressure. Valleys can also store great quantities of water and are the best place to store water in areas of low rainfall.

Layers of porous material below the earth surface often contain large amounts of confined or unconfined water, especially in areas with heavy seasonal rainfall.

There are many factors that influence the development of water in the alluvial stratums that lie below the earth, such as depth to the permeable zones, thickness of the stratums, width of the stratums, and permeability of the stratums.

Where there are natural barriers, such as vertical rock or clay formations to confine the water, a natural underground reservoir results. The water may be extracted from the reservoir by drilling a well and installing a pumping system. It is best to look for an underground water source during the dry time of the year, as possible oasis places will be most evident at that time. Underground water supplies may be found by checking vegetation (see water vegetation list in Chapter I), and geological formations.

Where there are no barriers, the water may be flowing in underground channels or it may be a shallow sheet flow across bedrock or some other impervious formation. You may be able to confine the water by creating artificial barriers or restrictions. If you construct your water storage facilities in a valley, a small backhoe may be used to cut trenches across the valley. Pack the trench with a good heavy clay, concrete, or other nonporous material. The trench must be deep enough to reach bedrock or a heavy clay subsoil. It will probably be necessary to pump the water from this reservoir, as the water will be stored underground.

If the reservoir is located at a higher elevation than the point of use, you can have gravity-flow pressure. It will be necessary to install a discharge pipe at the time the trench is dug and before the nonporous-material fill is put in place. The discharge pipe is installed by digging a trench at a right angle to the barrier trench and about 6 inches shallower than the deepest part of the barrier trench. Attach screen to discharge pipe and lay in trench. Cover the screen end of the pipe with about 1 foot of gravel and proceed to fill in both trenches with the nonporous-material fill.

If the water flow is deep, you may construct underground barriers by using drilling equipment. With drilling equipment, barriers can be constructed to almost any depth. When a drilling rig is used, holes are drilled at close intervals across the width of the valley to the bottom of the porous material that is to be used for the reservoir. Concrete is then pumped into these holes under pressure to create the artificial barriers. If a small area is left permeable near the bottom of the barriers, water can seep through and migrate to a point of discharge farther down the valley. The water stored behind the barriers continually replaces the water lost through discharge, so there is a constant flow of water from the point of discharge. A series of these barriers can be constructed to turn an entire valley into a series of vast underground water reservoirs.

If the valley fill is of the heavy clay-type material, it may be necessary to build an open-type reservoir.

This type of reservoir can be dug with a bulldozer. If there is a stream in the valley, the reservoir should be located to one side of the stream. If it is more practical to locate the reservoir in the stream bed, there should be a bypass constructed around the reservoir to carry flood waters away. If this is not done, the reservoir will soon fill with silt and debris.

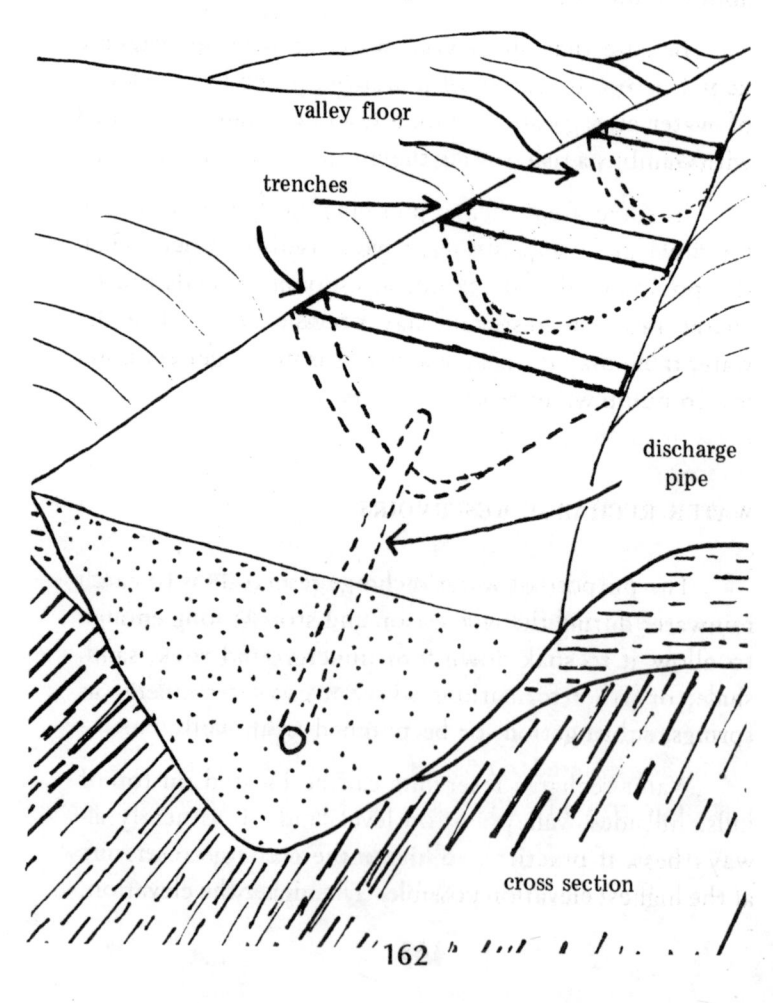

If you construct your water storage facilities on a hilltop, a bulldozer can be used to scoop out a basin. Be sure that the basin is deep enough to hold all of the water that falls in a year. If you live in an area that has a maximum rainfall of 80 inches, the basin should be at least 80 inches deep. If the soil is not a clay type that seals off water, a six-inch layer of pure clay soil should be spread over the basin and well packed. The area of the basin should be large enough to supply all of your water needs. To compute the size of the basin in square feet, take the minimum annual rainfall in feet x 7½. This will give you gallons per square foot. Take the gallons needed per year and divide by the gallons per foot. This will be the area needed in square feet. It is better to build the reservoir large enough so there is an excess above actual needs.

Be sure that all of your water storage facilities are kept free of all sources of pollution. Keep livestock out of water storage areas. Open reservoirs can be stocked with Gambusia fish to keep them free of mosquito larvae.

In some areas, heavy pumping has lowered the water table and some springs have ceased to flow. When the pumping of water from a reservoir exceeds the recharge rate, the reservoir may become depleted or the water table may drop so low that it is no longer economical to pump water from the reservoir.

WATER RECHARGE RESERVOIRS

The purpose of water recharge reservoirs is to catch rainwater during the wet season and store it long enough to allow it to soak down into underground rock, sand, shale, or other formations where it can later emerge as springs, subirrigation, or be pumped from wells.

Water recharge reservoirs can be located on top of hills, hillsides, valleys, or on level land. It is nearly always best, if practical, to locate the recharge reservoirs at the highest elevation possible. The higher the elevation, the longer it will take the water to percolate to the point of discharge, thus more water can be stored for a longer period of time. This is a very important consideration in areas that have wet seasons followed by long hot dry seasons.

The recharge reservoir should be deep enough to hold all of the water it will receive during the wet season. If the average rainfall is 50 inches during the wet season, the reservoir must be at least 50 inches deep. It is always better to allow some extra depth in case of above-normal rainfall. If the soil in the reservoir is too porous, allowing the water to drain too quickly, the reservoir should be partially sealed with clay. This will allow the water to remain in the reservoir for a longer period of time.

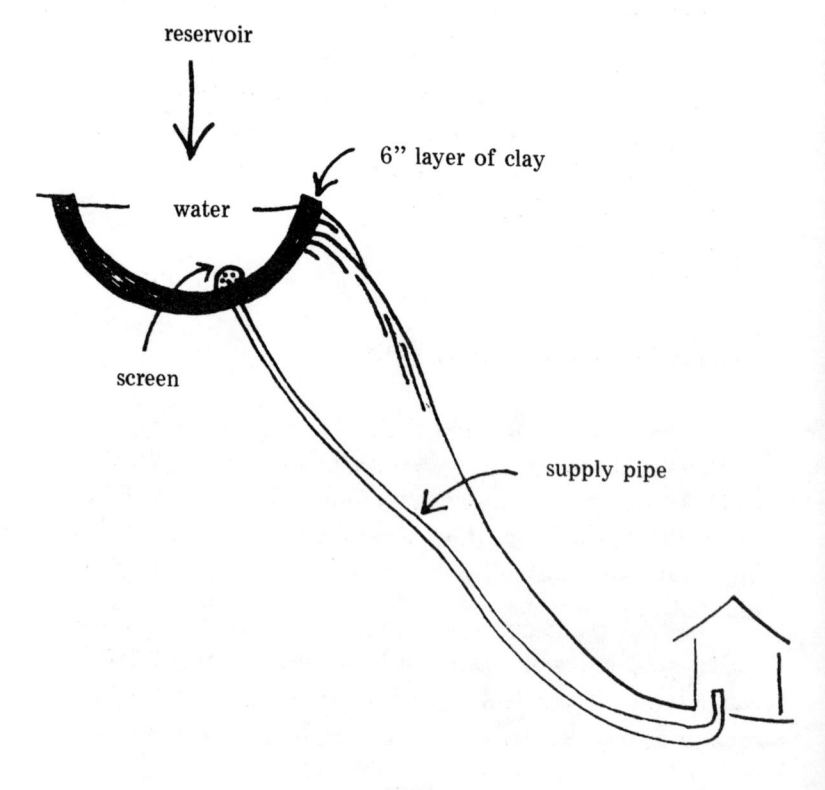

DAMS

Dams may be built of earth, concrete, or rock. Dams are generally built for water storage but can also be a source of electricity. Water stored behind a dam can be run through a turbine or water wheel that is connected to a generator, on its way to an irrigation ditch. Thus, it serves the dual purpose of irrigating crops and furnishing electric power. Dams are very successful if constructed on a clear, flowing stream that has a good year-round flow, but with modifications they can be successful on almost any stream.

EARTH DAM

An earth dam is the most economical dam if the material at hand is stable and will pack into a tight, leakproof structure. Be sure to test the material before using it in a dam. Take some of the material and build, on level ground, a small bowl-like reservoir 2 or 3 feet in diameter and about 1 foot high. Be sure that the material is moist enough that it will pack well. Fill the reservoir with water and let it stand. If the water seeps out in 20 or 30 minutes, the material is probably too porous for a dam unless a clay sealer is used. If the material dissolves or becomes soft and spongy, it is unstable and should not be used for a dam.

When building an earth dam, be sure to build a spillway around the dam. The spillway must be able to carry all of the stream flow plus flood water. It is very important that the spillway carry all of the water so that none of the water runs over the top of the dam.

The dam should be at least 3 feet above the highest water level. If the bottom of the spillway is not of bedrock or some other solid material, it should be concreted.

The inside wall of the dam should not be too steep, because a steep wall is hard to seal if a leak should develop. A steep wall is not as strong as a gently sloping wall, and it is subject to slides and erosion. A one-to-three slope is very satisfactory and a good rule to follow. A one-to-three slope means that you build outward 3 feet horizontally for each foot of elevation of the dam.

Example: (As shown in the diagram.)
A 10-foot-high dam would extend horizontally 30 feet up stream.

This method may also be used to set up a series of dams along a creek to provide a better flow of water year-round.

Boulders can also be used for constructing a series of dams. These dams may be built rather loosely when in series, allowing some of the water to flow through.

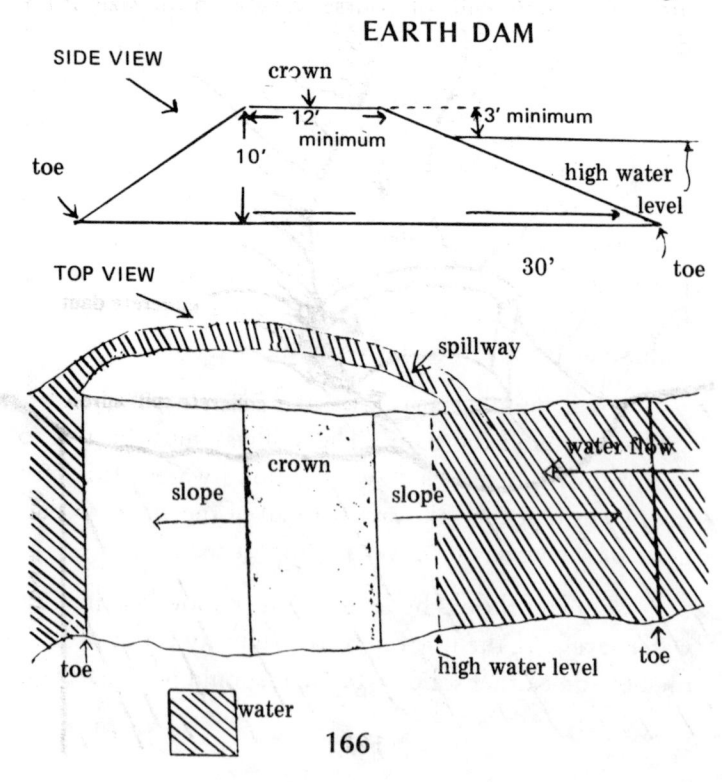

EARTH DAM

CONCRETE DAMS

A properly constructed concrete dam is very durable and should last indefinitely. A concrete dam can be very beautiful and it is not necessary to build a spillway around the dam because water running over it does not wash it away, as is the case with an earthen dam.

The first step in constructing a concrete dam is to dig a trench for the footing. The trench should be about 2 feet wide or more, depending on the size of the dam, and should be deep enough to reach a solid nonporous ground. The trench and dam should curve inward against the flow of the water. This type of dam is much stronger than a straight dam.

The footing and the dam should be reinforced with both horizontal and vertical steel bars. The size and the amount of bars will, of course, vary with the size of the dam.

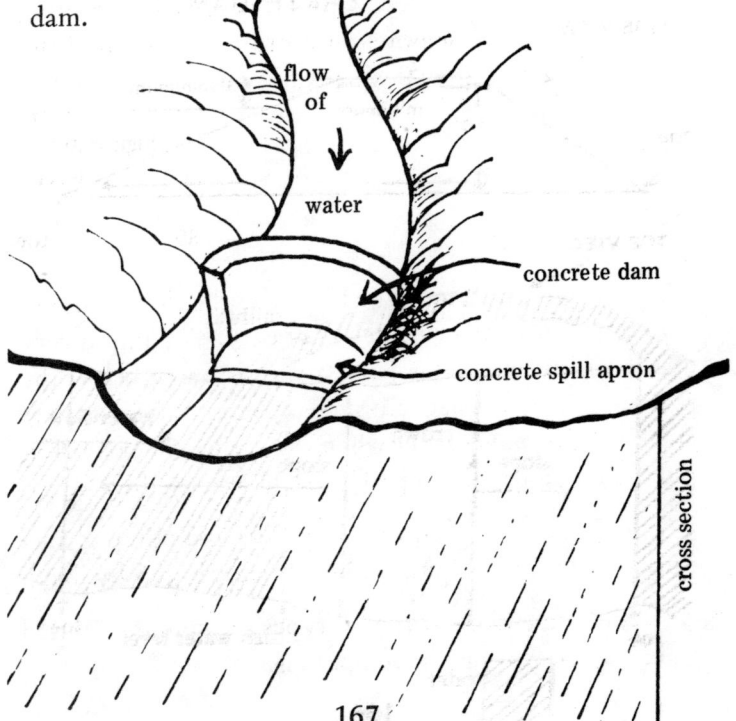

After the footing has been poured and has set one day, the rest of the dam may be completed. You can build a form and pour the dam in one piece, or it may be poured in increments 1 or 2 feet in height.

The concrete may also be placed on the footing with a shovel and formed by hand. The latter makes a very beautiful, handcrafted dam, and is much easier and less expensive to build than preparing forms for the concrete dam. About 6 inches height can be worked at a time. Dams should be built during the dry season, as there is less water flow and drying time is shortened. If the dam is long enough, by the time you have worked from one end to the other the starting point may be set hard enough that you can begin another 6-inch course of the dam. A 1-2-3 concrete mix should be used, that is, one part cement, two parts sand, and three parts gravel.

Keep the concrete moist for about 10 days by spraying with water several times a day, or by covering with a damp cloth. The center portion of the top of the dam must be low enough that water does not go around the ends of the dam.

PUMPS

A pump is a device which induces the flow of liquid against some resistance, such as gravity. A pump is very useful for raising water from a water source below the point at which it is needed, such as for extracting spring, creek, well or pond water when the spring, creek, well or pond cannot be located to provide gravity-flow water.

The simplest form of pump is a hand pump (sometimes called pitcher pump), which is operated by hand.

Electric pumps of various types are commonly used today, as are gasoline-driven pumps.

HOW TO BUY A WATER PUMP

1. Make sure that it is the type of pump for the job.
2. Select a pump with a large enough capacity for all your needs.
3. Be sure that the head pressure is high enough for all your needs.
4. Always buy from an established and reliable pump dealer.
5. Make sure parts and service will be available for the brand of pump that you select.
6. If you are buying an electric pump, make sure that the electrical requirements of the pump are available at your location.

HOW TO INSTALL A PUMP

1. Do not use a smaller suction pipe than the suction opening of the pump.
2. If the suction pipe is to have a long horizontal run, use a suction pipe one size larger than the suction opening of the pump.
3. Horizontal runs of suction pipe should always slope downward away from the pump toward the water source. This is to prevent air locks in the pipe.
4. Make sure that there are no leaks in the suction pipe. This is one of the greatest causes of pump failure.
5. If the discharge pipe is long, be sure to use a pipe large enough to prevent undue friction loss.

6. Make sure that pump is full of water before starting the motor (or before hand pumping). This is called "priming the pump". Many new pumps are damaged by running the pump while it is dry.

7. Build a pumphouse that will protect the pump from freezing and keep the electrical controls clean and dry.

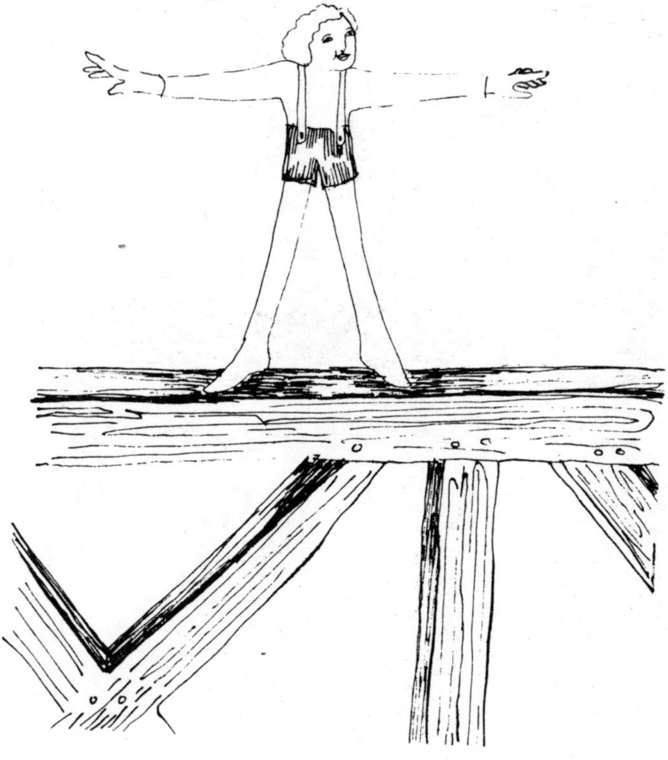

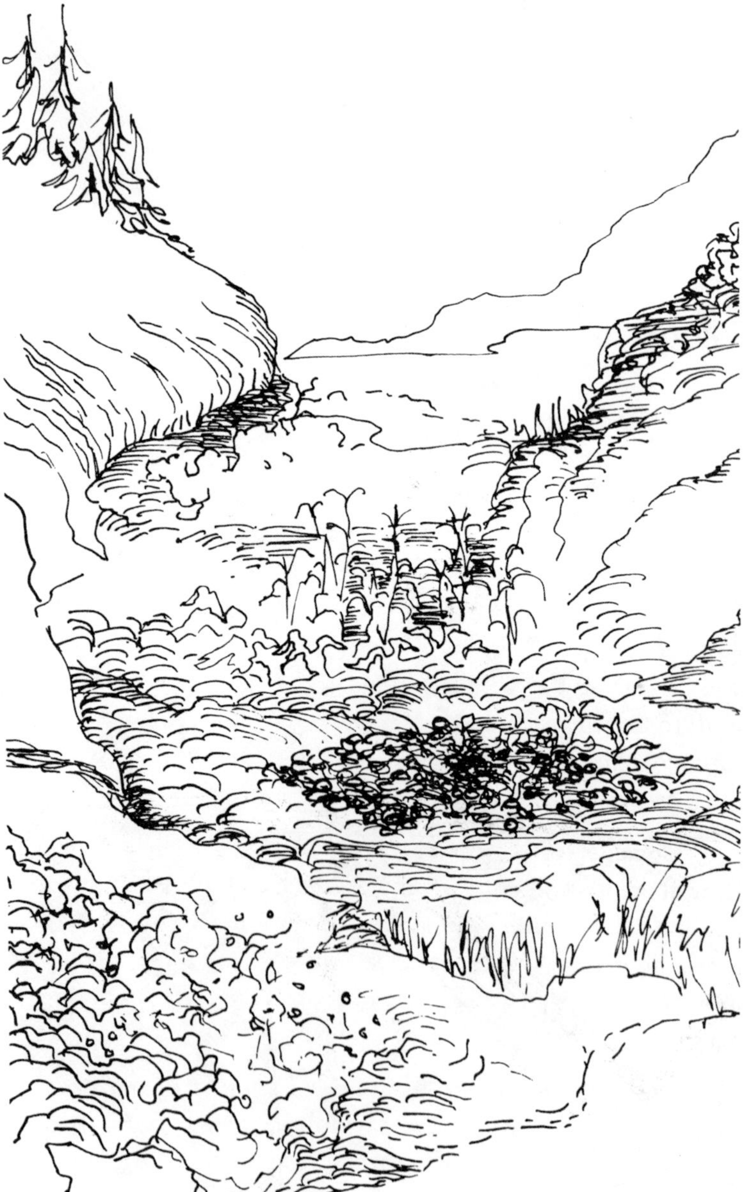

CHAPTER IV
ORGANIC GARDENING

CHOOSING LOCATION OF THE GARDEN

First of all, consider the direction of slope. As already discussed in the first chapter, a south or east slope is best for a garden, where early morning sun and midwinter sun will reach it.

Consider the path of the sun during the different times of year. The sun traces an arc in its path across the sky. During the summer, this arc goes directly east to west. During the winter, the arc follows a more southeast to southwest pattern. Be sure to take this change into consideration if you are planting the garden in a hilly area or an area surrounded by trees.

Consider the access to water for irrigation. Gravity flow from a spring or reservoir is ideal, unless you enjoy hauling water. A pump system from a reservoir lower than the garden would be an alternative, but this takes more energy (time, finances, and effort) to set up and maintain.

Consider natural or future development of microclimates. It is good to have a windbreak, such as hill or trees on the storm side of a garden, provided it does not shade the area. Planning the garden near a pond will help to temper the climate and possibly lengthen growing season.

Plan the garden area according to your personal needs and tastes, and local terrain. If land is in hilly country, you may want to plan a terraced garden. A terraced garden conserves space and irrigation water and can be very beautiful.

Consider the space between house and garden. Generally, the closer it is to the house, the better you will be able to watch over it and deter unwelcome visitors such as deer and ground squirrels. The trip between garden and kitchen will also be shortened, conserving valuable minerals from the vegetables, as well as valuable time in tending the garden.

Examine the soil in the various places that you are considering putting the garden. Soil richness and texture often vary, even within the space of a small area. Remember, though, that all soils can be improved, and that through conscious effort and maintaining a careful balance, your garden will improve more and more each year.

THE FENCE

A fence may be necessary to keep out such unwanted vegetarian creatures as deer, squirrels, gophers, and livestock. Most rodents and livestock will be discouraged by a well built 4-foot-high fence, but to keep out deer an 8 or 9-foot-high fence will be needed.

The most practical sort of fence to build for a garden is one using wood, concrete or metal posts placed at intervals around the garden space. Wire, chicken mesh, barbed wire or a combination of these is then strung around the posts and attached to them. Aluminum wire or wire made of some other shiny substance is often used, as animals such as deer will see it and not attempt to jump it.

Hedges, thick berry vines or stone walls may be used for part of the fence, but be certain that the garden will not be shaded by them. Also be sure not to leave spaces between this wall and the wire fence where animals may get through. The second chapter of this book has a section on fence construction.

To keep out burrowing rodents, such as gophers, bury fine-mesh chicken wire 2 or 3 feet deep around the perimeter of the garden, in line with the fence or as a continuation (downward) of it.

To keep out climbing rodents, attach a 3-foot-wide piece of fine wire mesh to the bottom of the fence around the perimeter. Bend the top foot-and-a-half outward at a 45 degree angle from fence. The climbing rodent will try to climb the fence but find himself falling back to the ground instead.

IRRIGATION WATER SYSTEM

In areas of irregular or sparse rainfall, irrigation will be necessary to carry water to the garden.

Run water lines from the pond or water storage tank to the garden with a shut-off valve at the garden site. These pipelines may be of either metal or plastic. In areas of severe frosts and cold weather, pipes should be buried underground a depth of 1 or 1½ feet. This will prevent the water in the pipe from freezing, causing pipe to crack or break, cutting off water supply to the garden.

Plan the method of watering you will use:

1. Hose placed individually at each plant.
2. Perforated hose running along rows of plants.
3. Rotating sprinkler on ground.
4. Overhead sprinkler.
5. Underground water irrigation.
6. Canals between rows running at a slight slope downward.

COMPOST

The compost bin plays a vital part in maintaining the balance of an organic garden. Plant and animal wastes are returned to the earth and are mysteriously transformed into rich humus containing all the minerals needed for healthy plant growth. Composting of raw wastes can take anywhere from a few weeks to a few months to complete, depending on climate, method of composting, and material to be composted.

COMPOSTING DIRECTLY

The simplest method of composting is to place organic wastes directly into garden spot. This method works best with a double garden site, one for spring planting, one for late planting. While one garden is growing, the other site may be composting. Additional benefit is derived from letting chickens graze in the composting gardensite. They will help disintegrate the organic matter, turn the earth with their scratching and pecking, rid the location of harmful insects, and add valuable manure to the future garden. The only drawback with this form of composting may be the odor of uncovered compost, and the fact that it may draw flies.

A very good garden arrangement is to have two adjacent gardens with the chicken house on the north and arranged so the chickens can be turned into either garden. This arrangement has several advantages.

The chicken house acts as a windbreak, protecting the garden from cold north winds and makes it easy to let chickens in the garden to clean up the remains from the harvest. This not only cleans up garden waste but turns it into rich humus and adds valuable manure to the soil. Compost (raw) may be dumped directly into the garden site, also, for the chickens to work with. The chickens will clean up the grass and weeds, leaving the ground in good shape for planting the next garden crop. While one garden is growing food, the other garden is being worked and fertilized by the chickens.

Another simple method of direct composting is to dig a pit in some vacant spot in the garden large enough to accomodate the compost that you empty from your indoor compost container. Each time you empty it, dig a pit in a new spot, and before you know it the entire garden will have an underground compost reservoir, continuously carrying nutrients to the plants. After the pit is dug and compost emptied, cover with earth.

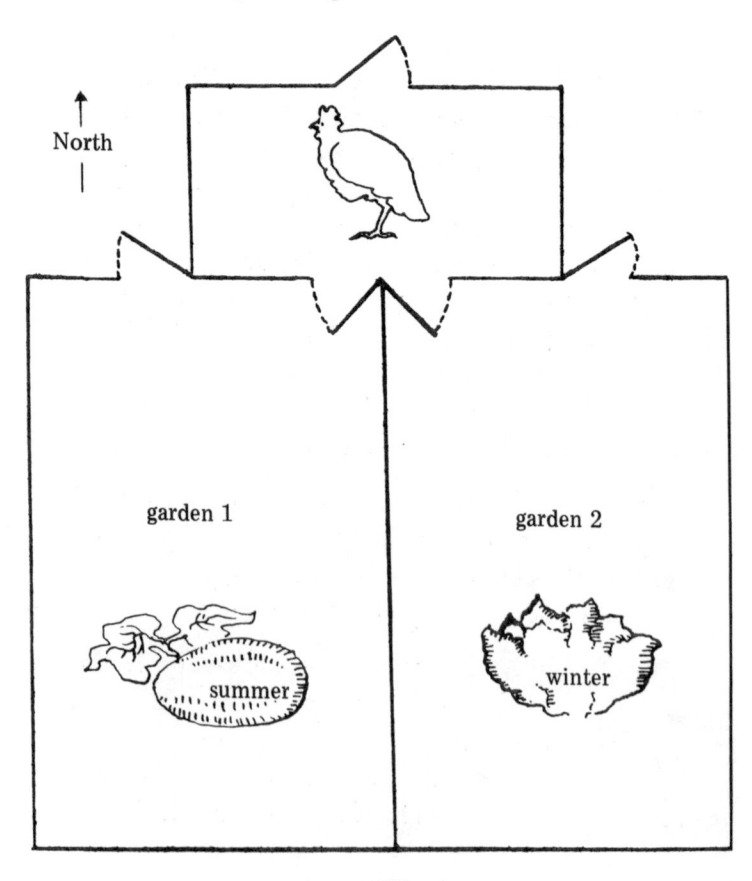

↑ North

garden 1   garden 2

summer   winter

COMPOST BINS

These may be built as simply or as elaborately as you wish. Important factors to consider:

1. Material used for bin construction.

    Wood: Easy to work with, but decomposes rapidly when used to confine compost, because of high heat and energy transfer.

    Metal: New or used metal cans and large containers may be easily procured, but metal also tends to decompose rapidly.

    Concrete: The most durable of all the materials and fairly easy to work with.

    Stone: Also very durable and easy to work with. Esthetically pleasing. May often be found free.

2. Ease in loading raw wastes and gathering finished compost.

3. Air Circulation: Compost needs air circulation for decomposition. The best way to provide for this is to have holes in the sides or bottom of bin for air to enter and circulate freely.

4. Ability to rotate composting materials. Rotation increases air circulation, which greatly shortens compost manufacturing time.

It is good to build the bin so that any liquid that drains through the pile may be saved and used as a liquid fertilizer. This liquid, formed by rain, watering the pile, or vegetable juices is extremely rich in minerals and should not be allowed to leach away.

COMPOST BINS

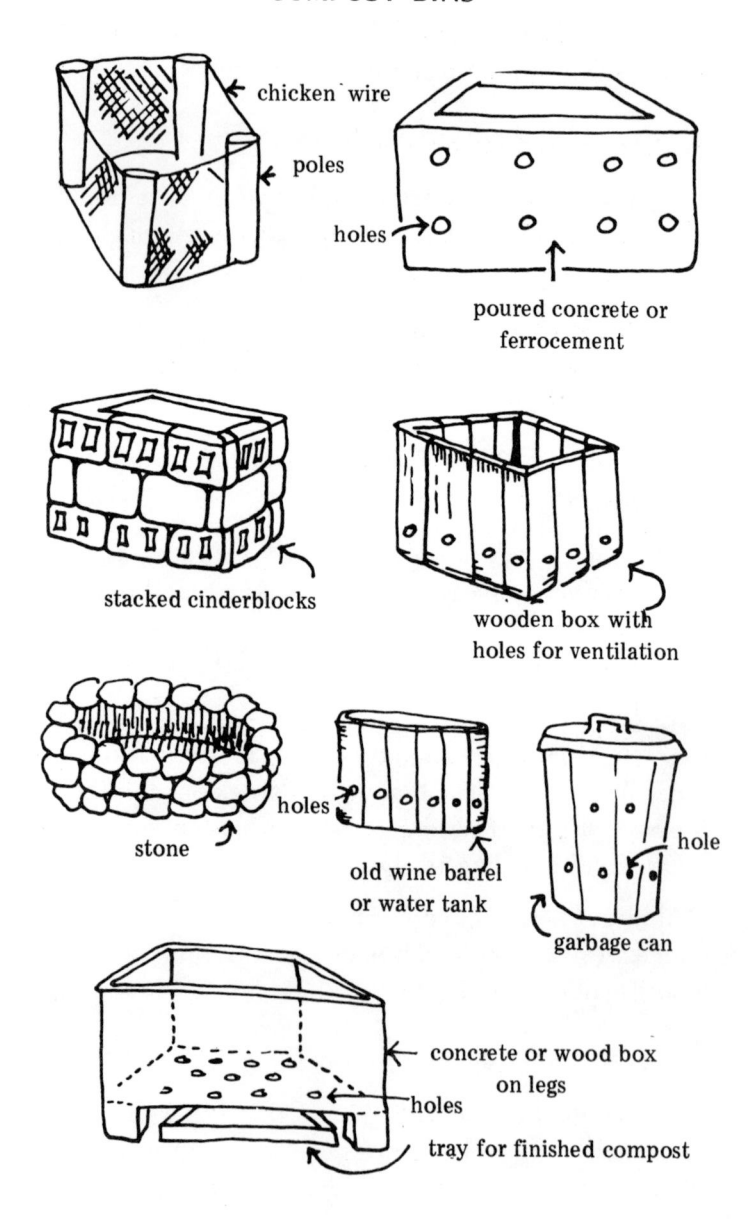

HASTENING COMPOST ACTION

1. Layering the various materials helps each to decompose quickly. No special order is necessary:

   soil
   organic wastes
   soil
   manure
   coarse material

   bonemeal
   phosphate rock
   bonemeal
   leaves

2. Aerating the compost by means of ventilation holes and turning materials with a pitchfork.

3. Watering compost, just enough to keep pile moist.

4. Addition of the following to the compost pile:

   manure
   bonemeal
   leaves
   feathers
   aquatic plants
   any plant or mineral rich in nitrogen

   soil
   kelp
   grass
   wood ashes

   earthworms
   bloodmeal
   milk products
   fish scraps

It is best not to add:

   diseased plants
   insect-infested plants
   chemicals of any kind
   wastes from vegetables grown with pesticides

Begin a new compost pile each winter, as this will allow the old one to mature and be ready for the spring. For this purpose it is often best to have two bins, so that one may be maturing while the other is being filled.

PREPARING THE SOIL TEXTURE

In order for plants to grow, texture of the soil must allow for aeration, drainage, and unhampered root growth.

Aeration is extremely important. Air spaces between particles of soil give plants oxygen and help support the soil bacteria that help assimilate nutrients used by plants. Good drainage is essential, as it provides for proper aeration. Soil must also be tight enough to prevent valuable water and nutrients from leaching out. If your soil is not of perfect texture, a variety of things may be done to make it so.

TILLING THE SOIL

This breaks up the soil surface, aerates the soil, and allows weeds to be used as a sort of green manure. This may be done, using a digging fork, a spade, a shovel, a roto tiller and/or a tractor.

Clay-type or other fine-textured soils should be tilled when just slightly moist. If tilled when very wet, soil will pack and form hard lumps. If you try to till when this soil is too dry, it will be nearly impossible to break the surface.

ADDING AERATORS

For clay-type and other nonporous soils, the addition of the following materials will make the soil more porous:

compost (raw and humus)   leaves, grass and hay
sand and silt   small stones)
coal ashes

ADDING SOIL BINDERS

For sandy-type and other very porous soils the

addition of the following materials will make the soil less porous:

compost humus   milk products
clay   manure
leaves, grass and hay
plant a cover crop such as rye and turn it into the soil

Do not worry about the presence of a few rocks in the soil, as they help to aerate the soil and add valuable minerals.

TESTING THE SOIL

Soil should be tested for pH, as well as for available minerals. This may be done either with a soil test kit or by sending a soil sample to a laboratory. Once again, local vegetation can also tell much about the quality of the soil.

The pH factor is the acidity, neutrality or alkalinity of the soil. These conditions release certain nutrients in the soil and make them available for the plants to use. Various plants need various nutrients. For instance, those that grow in acid soil require nutrients released in an acid-type soil.

To make a simple test for acidity or alkalinity, obtain some litmus paper from the drug store. Take a sample of moist earth and insert a strip of litmus paper.

If blue paper turns red, it indicates acid soil.
If red paper turns blue, it indicates alkaline soil.
If either paper turns purple, it indicates neutral soil.

The following is a list of both wild and cultivated plants that prefer a high-acid soil.

azaleas	lady's-slipper	orchid
berries	lilies	peanut
butterfly weed	lupine	pecan trees
dutchman's breeches	magnolia	potato
evergreens	madrone	radish
ferns	manzanita	redwood trees
flax	marigold	spicebush
heath	mayflower	spruce trees
heather	moss	watermelon
hickory	mountain laurel	wintergreen
huckleberry	oak trees	yew trees

The following is a list of wild and cultivated plants that prefer low-acid soil:

apple tree	kale	pumpkin
barley	millet	rice
beans	mustard	rye
buckwheat	oats	soybean
cherry tree	parsley	squash
corn	parsnip	strawberry
eggplant	pea	tomato
endive	peach tree	turnips
gooseberry	pear tree	vetch
grape	pepper	wheat

The following is a list of wild and cultivated plants that prefer neutral to alkaline soil:

alfalfa	carrot	lettuce
asparagus	cauliflower	okra
beet	celery	onion
broccoli	clover	quince
Brussels sprouts	cucumber	spinach
cabbage	iris	Swiss chard
cantaloupe	leek	

IMPROVING SOIL FERTILITY

Once soil has been tested, you will know what nutrients the soil is lacking. It is best to add these just before planting. It is wise to add fertilizer (organic only!) to soil after the rainy season ends, if possible. Added during or just before will allow nutrients to leach out quickly during the seasonal rain.

Compost Humus added to the soil is a certain cure for any mineral deficiency. It also helps balance too acid or too alkaline soil.

Leaves increase acidity, add nitrogen, zinc, boron, calcium, and magnesium.

Limestone increases alkalinity, adds magnesium and calcium.

Fish Emulsion adds nitrogen and phosphorus.

Manure adds nitrogen and phosphorus.

Bloodmeal adds nitrogen.

Bonemeal adds nitrogen and phosphorus.

Legume Cover Crop adds nitrogen.

Dolomite adds magnesium and calcium.

Phosphate Rock adds magnesium, phosphorus, boron, calcium, and iron.

Wood Ashes add potassium, calcium, phosphorus, and increase alkalinity of soil.

Seaweed adds nitrogen and potassium.

Citrus Peels add potasium.

Sawdust increases acidity, adds zinc and boron.

Hair and Feathers add nitrogen.

Hulls and Shells add potassium and nitrogen.

PLANNING THE GARDEN

Draw a diagram on paper first to plan where you will plant. Take care to leave walking passageways between rows or groups of vegetables. When preparing the plan for the garden, remember that it is important to plant tall-growing vegetables on the north side of short-growing vegetables, so as not to shade them. There are some plants, however, that enjoy shade and may be planted in the shade of a tall-growing plant.

Here are some size comparisons:

SHORT	MEDIUM	TALL
beets	artichoke	climbing beans
cabbage	asparagus	climbing squash
carrots	broccoli	corn
celery	Brussels sprouts	Jerusalem artichoke
chard	cauliflower	peas
collards	eggplant	sunflowers
cucumbers	peppers	
endive	soybeans	
garlic	tomatoes	
kale		
kohlrabi		
lettuce		
melons		
mustard		
onions and leeks		
parsley		
parsnips		
peanuts		
potatoes		
radish		
rhubarb		
rutabaga		
spinach		
squash		
strawberry		
turnips		

COMPANION PLANTING

Some vegetables have an affinity for one another and seem to do well when planted near each other. Here are a few:

beans—marigolds (aromatic variety)
broccoli—nasturtium
cabbage—mint
cabbage—nasturtium
cabbage—tomatoes
carrots—peas
cauliflower—nasturtium
chamomile—everything
chives—broccoli
cucumbers—corn
garlic—everything except legumes
leeks—celery
lettuce—tall growing plants for shade
onions—beets
potatoes—green beans
radish—chervil
radish—lettuce
soybeans—corn
spinach—tall growing plants for shade
tomatoes—asparagus
tomatoes—marigolds (aromatic variety)

Here are some insect-repellent flowers to plant in the garden to add color while they keep the bugs away:

anise, aster, basil
chrysanthemum,
coriander, marigold
cosmos, other aromatic herbs
mint, nasturtium.

PLANTING BY THE MOON

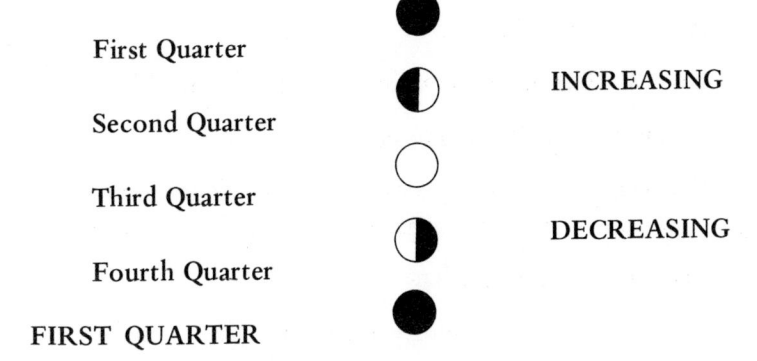

    First Quarter ●

    Second Quarter ◐   INCREASING

    Third Quarter ○

    Fourth Quarter ◑   DECREASING

FIRST QUARTER

Plant: asparagus, broccoli, Brussels sprouts, cabbage, cauliflower, celery, cereals, cucumbers, cress, endive, kohlrabi, lettuce, parsley, spinach, grains, watermelon, garlic, hay. Stimulates growth.

SECOND QUARTER

Plant: beans, eggplant, melons, peas, peppers, squash, tomatoes, grains, watermelon, garlic, hay. Stimulates growth.

THIRD QUARTER

Plant: bulbs, trees, shrubs, berries, beets, carrots, onions, strawberries, turnips, wheat, grapes. Pick fruit. Cut timber.

FOURTH QUARTER

Till soil, pull weeds, destroy pests, pick fruit. Retards growth. Cut timber.

THE ZODIAC SIGNS

Aries: Fire. Till soil, weed garden, plant onions.

Taurus: Earth. Plant bulbs and root crops.

Gemini: Air. Weed garden.

Cancer: Water. Very productive for planting everything, transplanting, and irrigating.

Leo: Fire. Weed garden.

Virgo: Earth. Till soil, weed garden.

Libra: Air. Plant hay, corn, grains, flowers.

Scorpio: Water. Productive for planting everything, transplanting and irrigating.

Saggittarius: Fire. Plant onions, grain, weed garden.

Capricorn: Earth. Plant bulbs, potatoes, and root crops.

Aquarius: Air. Till soil, eliminate garden pests.

Pisces: Water. Very productive for planting everything, transplanting, and irrigating.

PLANTING THE SEEDS

It is usually wise to soak the seeds overnight, especially for such plants as beets, chard, carrots, parsnips and onions, which germinate slowly. Soaking overnight hastens germination.

1. Place seeds to be sown in a jar of warm water, using a separate jar for each kind of seed.
2. Be sure to label each jar so you know what you have when you go to plant them.
3. Allow seeds to soak overnight or for 12-24 hours.
4. Drain on paper towel or other absorbent paper and allow to air dry for a few hours.
5. Plant directly into soil or mix seeds with sand if they are very small and sprinkle sand-seed mixture in the row.
6. Seeds may be sprouted before planting by soaking them overnight, wrapping in a paper towel, and placing in a cool dark place for a few days, keeping towel continuously moist, but not saturated. When small shoots appear from each seed, they are ready to plant.

Prepare the seedbed by tilling the earth, raking it smooth, and digging small trenches between rows or around areas to be seeded. These areas, then, become mounds. Smooth sides of mounds to slope gently downward so that seeds won't wash out.

These mounds are generally 6 to 12 inches high and are beneficial because:

1. They allow the earth to be warmed from three sides.

2. They keep plants from drying out because trenches allow water flow to be absorbed into the earth near the roots.

3. They provide for good drainage.

These mounds should be low, flat-topped mounds with shallow furrows so that the seeds will not wash away.

Sprinkle the seeds on the top of the mound and cover with a layer of fine-grained humus, or press each seed firmly into the earth and cover with fine-grained soil, or hollow out a small trench and cover with soil, or mix seed (especially for very small seeds) with fine sand and sprinkle seed-sand mixture on top of each mound, then cover with a layer of humus. Always be sure that earth is pressed down firmly to cover seeds. Mulch and water lightly.

Each type of plant has a depth at which its seed should be planted to give the best chance of survival. Smaller seeds are generally planted more shallowly than large seeds. The soil of the garden determines planting depth also. Water evaporates from the topmost layer of soil very quickly. The deeper you go into the earth, the higher the water content. As a light loam does not retain water very well, plant the seeds a bit deeper if this is the type of soil you are working with, as the added moisture at greater depth will help the seed to germinate. In a heavy soil that retains moisture well, such as clay, plant the seeds a bit more shallowly, as the higher water content deeper down may drown or decay the seed. Plant more deeply in dry weather.

By saving your own seeds from healthy plants each year for next year's planting, you will produce better and better vegetables as each year's crop becomes more and more acclimated to your particular climate.

TRANSPLANTING

Transplanting is the moving of small seedlings from one spot to another. This is often done to get a head start for plants that require a long growing season in an area with a short growing season.

Though beginning seed indoors and transplanting later has its definite advantages, it is usually best, if possible, to start seeds directly in the garden. Planted directly in the garden, they generally develop into plants with a larger root system, which helps them to withstand periods of drought and minor nutritional deficiencies. They are generally much stronger, more disease-resistant, and weather-hardy, especially when grown from seeds produced by your last year's crop. Also, there is no danger of "transplant shock" or cramped roots.

However, some plants may not have a long enough growing season in some areas unless planting is begun indoors several weeks before the last expected frost of the season. The advantages of transplanting are:

1. To get a head start on the growing season.

2. To control conditions for young seedlings to lessen the chance of fungus and disease.

3. Close observation and protection of young seedlings from insects, rodents, and other animals.

4. To provide warm earth for rapid seed germination.

5. To protect seed and seedlings from too much rain, wind or frosts of the early spring season.

INDOOR FLATS

Seeds are often started in indoor flats, which are

open boxes 3 or more inches deep that have spaces or drainage holes on the bottom.

1. Fill box to 1 inch from top edge with a moderately rich loam containing one part sand, one part compost humus, and two parts topsoil.

2. Soak seeds overnight in warm water if they are slow to germinate (see chart) and allow to air dry for a few hours.

3. Sprinkle seed thinly over the soil surface in the flat.

4. Cover with a layer of soil mixture to the depth recommended for planting whatever it is you are sowing.

5. Cover with a thin layer of sphagnum moss to prevent damping-off disease.

6. Mist lightly to moisten soil and keep it continuously moist until seedlings sprout.

7. Place flats in a room that is moderately warm, except for those plants that require a cool germination atmosphere. Room should receive some indirect sunlight.

8. When seedlings are large enough (see chart), they are ready to be transplanted to the garden, provided frost season has passed for warm-weather plants.

9. Prepare garden furrows and fertilize (organically!) as mentioned previously in this chapter.

10. Do transplanting on an overcast day or in the late evening after the sun has set. Seedlings are often "shocked" by sudden change in atmosphere and are not as easily damaged when transplanted during a time of minimal sunlight. If you must

transplant on a sunny day, prepare small boxes to cover each seedling as it is set out.

11. Water flat thoroughly, as this also helps to lessen transplanting shock.

12. Carefully remove each seedling along with the earth that is still attached to the roots. Try to avoid root breakage, although it is not as damaging as it may seem. Plants will generally recover rapidly.

13. Place each seedling in a hole large enough to hold entire root system without cramping. Set plant slightly deeper into the earth than it was in the flat.

14. Fill in the space with soil mixture and press down firmly to settle the earth around the root system. If air spaces are left beneath the surface, roots will dry out and plant may be damaged.

15. Water lightly.

16. Mulch around the base of each plant to protect it from the shock of a new environment and to help keep moisture in.

INDIVIDUAL PEAT OR PAPER CUPS

These are useful for larger seeds, as they can be started indoors and set directly into the garden at transplanting time. Since root system remains undisturbed, transplanting shock is lessened.

THINNING

When plants, especially those started from small

seeds, are planted directly into the garden, they often sprout and grow too close together. This can be avoided by sowing seed more thinly in the beginning. If they still grow too close together, they may be thinned by picking the smallest, weakest seedlings to leave room for the largest, healthiest plants. Or they may be dug up with a small shovel and some of them moved to another part of the garden and spaced further apart. This may be done by following the same procedure mentioned earlier for indoor flats.

CUTTINGS

Many plants may be propagated by cuttings. These should first be rooted by placing the bottom end of the cutting (where cut was made) in a container of water or a container of very wet sand. Keep plant moist by spraying with mister or a sprinkler once or twice a day. Place cutting in a greenhouse or a shaded place while roots are developing. Plants that may be propagated this way are:

berry vines	honeysuckle	rose bushes
currants	ivy	sweet potatoes
fig trees	most houseplants	tomatoes
grapes		

Any plants which naturally produce "suckers", which are small stems springing from a bud on the roots or main branches, may be successfully propagated by cuttings.

GROWTH CHART

Vegetable	Days to Germinate	Weeks to Transplant	Months to Mature
artichoke	plant roots	4-6	5-6
asparagus	7-20	52	36
beans	6-14	3	2-3
beets	7-12	6	2-2½
broccoli	3-10	4-12	2-3
Brussels sprouts	4-12	8	4-5
cabbage	4-12	4-6	2-3
Chinese cabbage	4-12	4-6	2-3
cantaloupes	4-12	2-3	3-4
carrots	12-20	6	2-3
cauliflower	6-12	4-6	3-5
celery	12-20	8-9	5-6
celeriac	12-20	8-9	5-6
chard	6-12	3	1-2
collards	4-12	4-6	2-3
corn	6-12	3	3-4
cucumbers	5-10	3-4	2-3
eggplant	7-14	8-10	5
endive	6-10	4	2-3
garlic	5-12	4-6	3
Jerusalem artichokes	plant tuber	X	4
kale	4-12	4-6	2-2½
kohlrabi	4-12	4-6	2
leeks	6-12	4	3-5
lettuce	4-12	3-5	2-3
mushrooms	14-21	X	2-3
mustard	4-12	2	1-1½
okra	4-12	3	1½-2
onions	7-15	8	1-4
parsnips	15-25	4-6	3-4
peanuts	7-14	X	4-5

Vegetable	Days to Germinate	Weeks to Transplant	Months to Mature
peas	7-15	4-6	1-2
peppers	9-21	6-8	3-4
potatoes	8-15	8	3-4
potato, sweet	30	6-8	3-3½
radish	3-10	1-2	1-2
rhubarb	plant tubers	X	12-18
rice	30-60		4-5
rutabaga	4-12	4-6	3
sesame seeds	9-21	6-8	3-4
soybeans	6-12	4-6	2-3
spinach	7-14	2-3	1-2
spinach, New Zealand	7-12	4-6	2-3
squash, summer	3-12	3-4	2-2½
squash, winter	7-12	4	3-4
strawberries	plant runners	X	6-18
sunflower	7-12	4-6	3-3½
tomatoes	6-15	4-8	2-4
turnips	4-12	2	1-2
watermelons	3-15	6-7	2-4
wheat	3-12	X	3½

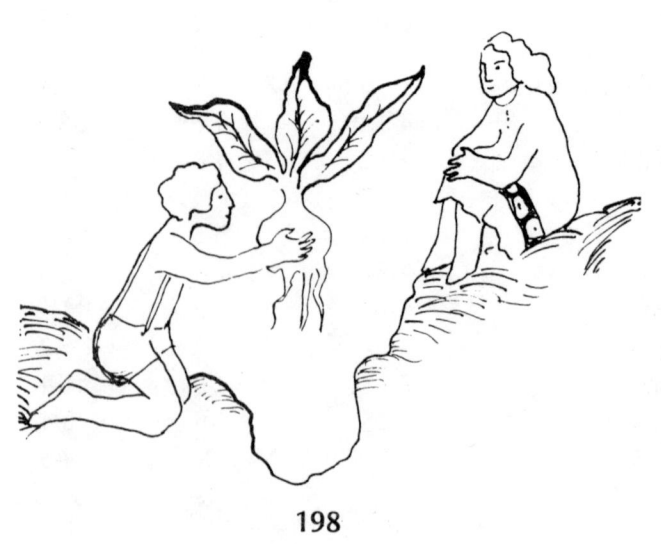

THE VEGETABLES

The following pages are a guide to the individual requirements for each vegetable, nutritional as well as environmental. They should be used as a guide, but not as a substitute for experiencing and "tuning in to" the needs of your garden.

The geographical divisions, which show when each plant may be safely set or sown outdoors, should be used as a basic guide and not taken too literally. All sections of the country are not uniform in climate, as air and water currents, microclimates, and many other factors, influence each valley and hilltop. Valleys, especially high valleys, generally frost earlier in the season, but temperatures are more moderate than hill or mountain tops, which are generally windier, more extreme in temperature, and receive snow earlier in the season.

ARTICHOKE

WHEN TO PLANT:

Northeast—April
Southeast—September-December
Midwest—April
Northwest—April
Southwest—September-January
Far North—April
Coastal—September-May

Soil: Deep, rich, well drained, composted soil; pH: 6-8 (low acid or alkaline to neutral). Nutritional Requirements: nitrogen.

Planting: Start from roots or suckers placed 6 inches deep directly into garden. Or start seeds 4-6 weeks early in indoor seedbeds. Space plants in garden 4-6 feet apart.

Temperature: Artichokes are Moderately Hardy and a Cool Weather plant. They thrive best in mild, cool, humid weather with indirect sunlight. Can stand frost if mulched.

Water: Continuously moist, but well drained. Do not allow crowns to collect moisture.

Seasonal Protection: Artichoke is a perennial and will continue to produce for 3-4 years if protected during frost season by placing a large basket over plant and covering it with a mulch of manure or hay. Mulch base of plant in summer to conserve moisture. Protect crowns from heavy rainfall.

Planting Seed to Harvest Time: 5-6 months.

Harvest: Gather thistle-like heads before they bloom; remove all before heavy frost arrives. Yield: 10-20 heads per plant.

ASPARAGUS

WHEN TO PLANT:

Northeast—April
Southeast—March
Midwest—April
Northwest—April
Southwest—March
Far North—April
Coastal—March

Soil: Rich, deep loam, not too porous; pH: 6-7 (slightly acid). Nutritional Requirements: nitrogen, lime.

Planting: Sow seeds ½-1 inch deep in flats or in out-of-the-way places in the garden. Should be kept warm and moist while seed germinates. When plants are 1-2 years old (or procure 1-2 year old roots) dig a trench 1 foot deep and place in it a mixture of well-rotted manure, compost, and limestone, and on top of that a layer of compost humus. Place 1-2 year old roots about 6 inches deep and fill in with topsoil. Space plants 1-2 feet apart.

Temperature: Asparagus is Hardy and a Cool Weather plant. It thrives best in a cool location with partial sun. Can stand frost if mulched.

Water: Continuously moist.

Seasonal Protection: Asparagus is a perennial. It should be mulched heavily at all times of the year, in summer to conserve moisture and in winter to protect from frost. Plants will produce for 10-15 years.

Planting Seed to Harvest Time: 3 years.

Harvest: Pick tender shoots beginning the third spring. At this time, harvest only a few shoots—about a dozen or so. The second harvest you may pick more, increasing yield each spring. Overpicking will cause slow development and low yield next year.

BEANS

WHEN TO PLANT:

Northeast—May-June
Southeast—April-August
Midwest—May-June
Northwest—May-June
Southwest—May-July
Far North—May
Coastal—April-July

Soil: Rich clay or sandy loam, not too heavy, well drained; pH: 5.5-6.5 (low acid). Nutritional Requirements: phosphorus and potash. Note: Beans add nitrogen to the soil in which they grow and there is no need to add a nitrogen supplement. In fact, the addition of nitrogen to their soil tends to make them produce profuse leaves and few or no pods.

Planting: Soak dried beans overnight. Sow in soil that is warm and moist, not cold and damp, as bean seeds decay rapidly under such conditions. Seeds may be started 3 weeks earlier in indoor seedbeds. Plant 1 inch deep, space 6-12 inches apart. At time of planting, prepare poles, trellis, fence or tripod for beans to climb on.

Temperature: Beans are Not Hardy and are a Warm Weather plant. Temperature must be 60-95 degrees and frost-free during growing season.

Water: Soil must be kept moist at all times.

Seasonal Protection: Protect from extreme heat and heavy rains. Mulch to preserve soil moisture.

Planting Seed to Harvest: 2-3 months.

Harvest: Pick edible-pod beans when young and tender. Let drying beans dry on plant, then pick and thresh.

BEETS

WHEN TO PLANT:
Northeast—April-July
Southeast—March-August
Midwest—April-July
Northwest—April-July
Southwest—September-January
Far North—April-July
Coastal—February-August

Soil: Deep sandy loam, well drained and stone-free; pH: 6-7.5 (neutral); will not grow in acid soil. Nutritional Requirements: lime.

Planting: Dig in a small amount of compost and lime into garden before planting. Do not fertilize too heavily, as this encourages top (leaf) growth rather than bulb growth. Soak seeds overnight. When planting directly into garden, sow thinly, as each beet seedpod actually contains several seeds. Plant 1 inch deep and space 5 inches apart. May be started 6 weeks early in an indoor seedbed. It is good to do a continuous planting every 1-2 weeks during planting period to have a continuous supply.

Temperature: Beets are Hardy and are a Cool Weather plant, but do not like shade, so plant in full sun.

Water: Very moist at all times.

Seasonal Protection: Keep well mulched to prevent them from becoming too dry and to protect them from intense heat.

Planting Seed to Harvest Time: 2-2½ months.

Harvest: Pull beets when 1-3 inches in diameter or mulch in fall and harvest all winter as needed.

BROCCOLI
WHEN TO PLANT:

Northeast—March & July
Southeast—March & August
Midwest—July
Northwest—March & July
Southwest—March & September
Far North—June
Coastal—June-September

Soil: Rich loam; pH: 5.5-6.5 (slightly acid-neutral). Nutritional Requirements: calcium.

Planting: Plant seeds ½ inch deep directly into garden or start seedlings 2-3 months early in indoor seedbeds. Space plants 2 feet apart.

Temperature: Broccoli is Very Hardy and a Cool Weather plant. It likes warm days and cool nights, but does not like extremely hot weather and should be placed in partial sun. A light frost improves the heads.

Water: Must be kept continuously moist.

Seasonal Protection: Mulch to prevent soil from drying out and shade from extreme heat. Mulch heavily in spring and fall to protect against sharp frost. Enjoys light frosts.

Planting Seed to Harvest Time: 2½ months.

Harvest: Cut flowering heads before flowers open. More edible flowering stalks will form.

BRUSSELS SPROUTS
WHEN TO PLANT:

> Northeast—May
> Southeast—June
> Midwest—April-May
> Northwest—April-May
> Southwest—July
> Far North—April
> Coastal—April-July

Soil: Rich, deep, sandy loam with good drainage; pH: 6-7 (slightly acid). Nutritional Requirements: nitrogen.

Planting: Except in areas with especially long growing seasons, it is best to start seeds in indoor seedbeds 2 months early. Space plants 2 feet apart.

Temperature: Brussels sprouts are Very Hardy and a Cool Weather plant. They enjoy light frosts. Plant in shade or indirect sunlight.

Water: Needs plenty of moisture at all times.

Seasonal Protection: Mulch to conserve moisture and nutrients. Keep shaded from too much sun. Prepare stakes for plants when they are 1 foot high. Remove lower leaves when buds sprout.

Planting Seed to Harvest: 4-5 months.

Harvest: Twist off firm sprouts before they become tough or yellow.

CABBAGE and CHINESE CABBAGE

WHEN TO PLANT:

    Northeast—July
    Southeast—February-August
    Midwest—July
    Northwest—July
    Southwest—July-November & Jan.-February
    Far North—July
    Coastal—September-January

Soil: Rich loam; pH: 6-7 (slightly acid). Nutritional Requirements: nitrogen, potash and magnesium.

Planting: May be sown directly into garden, or start seeds in indoor seedbeds in a cool place 4-6 weeks early. Space plants 2 feet apart.

Temperature: Cabbages are Very Hardy and a Cool Weather plant. Regular cabbages can withstand sharp frosts, but Chinese cabbages can withstand only light frosts. Both prefer a cool, moist climate in partial to full sunlight, but will bolt to seed if temperature goes too high.

Water: Very moist. Keep evenly damp at all times and do not allow uneven watering, that is, periods of dryness followed by drenching.

Seasonal Protection: Protect from drastic changes of climate, such as a long period of freezing weather suddenly changing to warm weather. Mulch heavily. Protect from dry winds.

Planting Seed to Harvest Time: 2½-3 months.

Harvest: Cut off from base as soon as heads are firm.

CANTALOUPES [small cantaloupes, honeydew, casaba, crenshaw and others of the cantaloupe family]

WHEN TO PLANT:

    Northeast—May
    Southeast—April-May
    Midwest—May
    Northwest—May
    Southwest—April-June
    Far North—May
    Coastal—April-June

Soil: Warm sandy loam; pH: 7-8 (neutral to alkaline). Nutritional Requirements: plenty of everything.

Planting: Plant seeds directly in garden after frost danger is past or start seeds in warm indoor seedbeds 2-3 weeks early. Sow seed ½ inch deep. Allow plenty of space in the garden, as they spread rapidly. They may be planted in groups of three, spaced 6 feet apart.

Temperature: Cantaloupes are Not Hardy and are a Warm Weather plant. They need full sunshine while growing.

Water: Abundant moisture until melons begin ripening.

Seasonal Protection: Mulch to conserve water and to protect fruit from touching damp earth. Protect from chilling winds.

Planting Seed to Harvest Time: 3-4½ months, depending on variety.

Harvest: A ripe melon, when picked up off the ground falls off the vine with no effort to pull it off. Smell, skin color and softness are other signs of ripeness.

CARROTS

WHEN TO PLANT:

Northeast—April-July
Southeast—March-July
Midwest—April-July
Northwest—March-July
Southwest—September-February
Far North—April-June
Coastal—January-August

Soil: Sandy loam; pH: 5.5-6.5 (slightly acid). Nutritional Requirements: plenty of compost humus. Do not use manure.

Planting: It is best to soak seeds overnight, as they are slow to germinate. They may also be sprouted first to hasten germination by sprinkling seeds between two layers of absorbent paper; keep continuously moist and store in a cool place until they begin to sprout. Sow seeds in the garden ½ inch deep or start them in indoor seedbeds 6 weeks early. Space 3-4 inches apart. Continual planting throughout growing season is best.

Temperature: Carrots are Very Hardy and a Cool Weather plant. They will survive sharp frosts but prefer an average temperature of 60-70 degrees and an open, sunny location.

Water: Need plenty of water. Irrigate during dry season.

Seasonal Protection: Mulch to prevent soil from drying out.

Planting Seed to Harvest Time: 2½ months.

Harvest: Pull roots when they are ½-1 inch wide across top.

CAULIFLOWER

WHEN TO PLANT:

Northeast—May
Southeast—March & July
Midwest—April
Northwest—April-May
Southwest—July-August
Far North—May
Coastal—February-March—August

Soil: Needs extremely fertile, moisture-holding yet well drained soil; pH: 6-7 (slightly acid). Nutritional Requirements: nitrogen, calcium, potash.

Planting: May be planted directly in garden, but it is usually best to start seed in indoor seedbeds 4-6 weeks early. Plant seeds ¼ inch deep. Space plants 2 feet apart.

Temperature: Cauliflower is Not Hardy and is a Cool Weather plant, though it will not withstand frost. It needs a moderate, even temperature and will not form a well developed head in extremely hot weather unless protected.

Water: Needs large quantities of moisture, especially when heads are forming.

Seasonal Protection: Very tender and must be treated gently. Will need to be shaded from too much sunlight. Protect when flower begins to form by tying its leaves up over the head or by providing another form of shade. Mulch heavily to conserve moisture and nutrients, and keep soil cool.

Planting Seed to Harvest Time: 3-5 months, depending on variety chosen and climate.

Harvest: Cut off flower as soon as solid head develops.

CELERY and CELERIAC

Note: Celeriac grows under similar conditions as celery but is more hardy in terms of moisture and nutritional requirements.

WHEN TO PLANT:
- Northeast—May-June
- Southeast—June-July
- Midwest—April-June
- Northwest—March-June
- Southwest—February-April
- Far North—April
- Coastal—March-August

Soil: Extremely rich, moisture-holding soil. Prefers a marsh environment; pH: 6-7.5 (slightly acid to neutral). Nutritional Requirements: great quantities of everything.

Planting: May be started directly in the ground in areas that have a long, cool, damp growing season, but it is usually wise to get a head start by planting seeds in indoor seedbeds 2 months early. Seeds, which are slow to germinate, should probably be soaked first and planted ¼ inch deep. Space plants 1 foot apart. Young plants should be shaded from direct sunlight.

Temperature: Celery and Celeriac are Not Hardy but thrive in Cool, Humid Weather. Temperature range should be between 50-70 degrees. Will not withstand frost.

Water: Plenty at all times to create a marshy environment.

Seasonal Protection: Mulch heavily to protect from light frosts and to conserve moisture.

Planting Seed to Harvest Time: 5-6 months.

Harvest: Cut at base for immediate use or pull out entire plant, roots and all, to store.

CHARD
WHEN TO PLANT:

Northeast—April-August
Southeast—August-May
Midwest—April-August
Northwest—April-August
Southwest—August-May
Far North—April-July
Coastal—Year 'round

Soil: Rich, well drained loam; pH: 6-7 (low acid to neutral). Nutritional Requirements: minimal.

Planting: Soak seeds overnight. Sow directly in garden 1 inch deep or start in indoor seedbeds 3 weeks early. Space plants 1 foot apart.

Temperature: Chard is Very Hardy and a Cool Weather plant. It thrives in both hot and cold weather.

Water: Moderate.

Seasonal Protection: Chard is a perennial and will continue to grow for 1-2 years or more before going to seed. Mulch in the summer to conserve moisture, and in the winter in very cold areas.

Planting Seed to Harvest Time: 1½-2 months.

Harvest: Cut leaves individually. Plant will continue to send up new leaves.

COLLARDS
WHEN TO PLANT:

Northeast—April—August
Southeast—March-July
Midwest—April-July
Northwest—April-June
Southwest—November-May
Far North—March-April
Coastal—September-May

Soil: Well drained loam; pH: 5.5-6.5 (slightly acid). Nutritional Requirements: moderately rich.

Planting: Seeds germinate quickly and may be sown directly in garden, or may be started in indoor seedbeds 4-6 weeks early. Sow seeds ½ inch deep and space plants 1½ feet apart.

Temperature: Collards are Very Hardy and a Cool Weather plant. They do well in direct sunlight, as well as overcast weather. Can withstand heavy frosts.

Water: Keep moist, as this will promote rapid growth which produces tender plants.

Seasonal Protection: Mulch to conserve water. A stake is useful to support stalk.

Planting Seed to Harvest Time: 2½-3 months.

Harvest: Pick outer leaves individually or harvest whole head at once.

CORN

WHEN TO PLANT:

Northeast—May-June
Southeast—April-July
Midwest—May-July
Northwest—April-May
Southwest—March-June
Far North—May
Coastal—April-June

Soil: Well drained rich loam is preferred, though it thrives in poorer soils as well; pH: 6-7 (slightly acid-neutral). Nutritional Requirements: nitrogen, calcium, potash.

Planting: May be planted directly into soil ½-1 inch deep, with successive plantings every week or so. Space plants 1-2 feet apart and plant in patches of 3 or more rows in all directions so that plants will pollinate each other. This is very important for the proper development of kernels.

Temperature: Corn is Moderately Hardy, but a Warm Weather plant. Withstands light frost, but prefers hot weather and direct sunlight.

Water: Needs plenty of water until after tasseling, then ears form more quickly with less water.

Seasonal Protection: Mulch heavily around base of plants. If plants need hand-pollination, stroke tassels that are heavily laden with pollen and place collected pollen on the cornsilk of each ear.

Planting Seed to Harvest Time: 3-4 months, depending on variety and climate.

Harvest: When cornsilk turns brown, kernels are evenly formed and a milky substance flows out of kernels when punctured, ears are ready to harvest. If, after silk turns brown, kernels are unevenly spaced and sparsely formed, it probably did not get pollinated. Popcorn or other drying corn is left on stalk until dry, then picked and stored.

CUCUMBERS

WHEN TO PLANT:
Northeast—May-June
Southeast—May-July
Midwest—May-June
Northwest—May-June
Southwest—March-August
Far North—May-June
Coastal—April-July

Soil: Well drained rich loam of equal parts sand, clay and humus. Should be extremely rich; pH: 5.5-6.9 (slightly acid). Nutritional Requirements: nitrogen.

Planting: Best to soak seeds overnight. Plant directly in garden or sow indoor seedbeds 3-4 weeks early. Sow seeds 1 inch deep, spacing about 12 inches apart. May be grown in either rows or hills. Hills are preferred where soil drainage is poor.

Temperature: Cucumbers are Not Hardy and are a Warm Weather plant. Ground should be warm at planting time and plants should be placed in a sunny spot. Will not survive frost.

Water: Constant supply of water is needed. Lack of water is often the main cause of bitter cucumbers. To supply sufficient water, place a can or clay pot, with very small holes punched in the sides near bottom, into earth in the center of each hill or between plants in a row. Each day, fill pot with water and it will seep out slowly, keeping ground continuously moist. Additional nutrients may also be supplied this way.

Seasonal Protection: Heavy mulching is necessary to keep ground continuously moist. Provide trellis, fence or tripod for cucumbers to climb on.

Planting Seed to Harvest Time: 2½-3 months.

Harvest: Pick when 1-2 inches long for pickling or when still young and tender for salads. Pick before they turn yellow.

EGGPLANT

WHEN TO PLANT:

Northeast—May
Southeast—May
Midwest—May
Northwest—May
Southwest—April-May
Far North—May
Coastal—April-May

Soil: Deep, rich, porous soil containing some sand; pH: 6-7 (slightly acid to neutral).

Planting: May be sown directly in garden in areas of long growing season, but it is sometimes wise to start seeds in indoor seedbeds 8-10 weeks early to shorten outdoor growing requirement, as eggplant is very tender and needs a warm growing season. Sow seeds ½ inch deep and space plants 2 feet apart.

Temperature: Eggplants are Very Tender and are a Warm Weather plant. The earth must be warm. They will not survive drastic temperature changes or frost. Temperature range outside should be 65-70 degrees, though some varieties are hardier than others.

Water: Needs much moisture, especially when fruit is forming.

Seasonal Protection: Mulch heavily to protect from chills, wind, and water loss.

Planting Seed to Harvest Time: 5 months.

Harvest: May be picked as soon as skin turns shiny or when more fully mature. Harvest before frost.

ENDIVE

WHEN TO PLANT:

Northeast—July-August
Southeast—August-March
Midwest—July-August
Northwest—July-August
Southwest—August-March
Far North—July
Coastal—August-March

Soil: Moderately rich loam; pH: 6-7 (slightly acid to neutral). Nutritional Requirements: phosphorus and calcium.

Planting: Sow directly in a shaded part of the garden or start in an indoor semi-shaded seedbed 4 weeks early. Make successive plantings every week or so. Sow seeds ½ inch deep and space plants 1 foot apart.

Temperature: Endive is Hardy and a Cool Weather plant. It is fairly heat resistant, as well as being frost resistant. Thrives best in a cool, humid atmosphere.

Water: Continuously moist.

Seasonal Protection: Mulch heavily to protect from frost as well as to conserve moisture. Shade from intense sun, especially when plants are very young.

Planting Seed to Harvest Time: 2-3 months.

Harvest: Use outer leaves as they develop, or cut at base to remove entire head. To store, remove entire plant, roots and all.

GARLIC

WHEN TO PLANT:

Northeast—April
Southeast—August-September
Midwest—April
Northwest—March
Southwest—August-September
Far North—April
Coastal—August-March

Soil: Moderately rich sandy loam; pH: 7-8 (slightly alkaline to neutral). Nutritional Requirements: nitrogen, phosphorus, potassium and trace minerals.

Planting: Plant the outside cloves of a garlic bulb or garlic sets 2 inches deep directly into garden, spacing them 6 inches apart.

Temperature: Garlic is Fairly Hardy and does well in both Cool and Warm weather, preferring mild temperatures during growing period.

Water: Keep moist during growing period but well drained. Do not water when bulb approaches maturity.

Seasonal Protection: Mulch to conserve moisture. Protect from heavy rains.

Planting Seed to Harvest Time: 3 months.

Harvest: Pull when tops bend to the ground and become dry. Dry bulbs in an arid space protected from direct sunlight.

JERUSALEM ARTICHOKES
WHEN TO PLANT:

Northeast—April—September
Southeast—March—September
Midwest—April—September
Northwest—March—September
Southwest—February—September
Far North—April—September
Coastal—February—September

Soil: Rich sandy loam; pH: 6-8 (close to neutral). Nutritional Requirements: A rich soil will improve size and yield.

Planting: Plant root tubers directly into garden 4 inches deep. Space plants 2½ feet apart. Roots spread rapidly and should be confined to keep from taking over.

Temperature: Jerusalem Artichokes are Very Hardy and a Cool Weather plant. They withstand sharp frosts but also thrive in full or partial sun.

Water: Abundant.

Seasonal Protection: Jerusalem Artichokes are perennial and will continue to produce as long as one or more tubers are left in the earth. Mulch heavily in the fall.

Planting Seed to Harvest Time: 4 months.

Harvest: Dig tubers in fall and all through winter. Cut stalks in fall to use as mulch.

KALE

WHEN TO PLANT:
Northeast—June-August
Southeast—August-October
Midwest—July-August
Northwest—June-August
Southwest—August-November
Far North—June-July
Coastal—September-March

Soil: Fertile sandy or clay loam; pH: 6-8 (slightly acid or alkaline to neutral). Nutritional Requirements: calcium, nitrogen.

Planting: Plant directly into garden ½ inch deep and space plants 6-12 inches apart. May also be started in indoor seedbeds 4-6 weeks early.

Temperature: Kale is Very Hardy and a Cool Weather plant. It can withstand frosts and is actually improved by light frosts. Not as flavorful when grown in hot weather.

Water: Abundant water is necessary to encourage rapid growth.

Seasonal Protection: Biennial, and will continue to produce for 2 years. Mulch to conserve moisture and protect from frosts.

Planting Seed to Harvest Time: 2-2½ months.

Harvest: Pick outside leaves before fully mature and inner leaves when still tender. This way, plant continues to grow and produce new leaves.

KOHLRABI

WHEN TO PLANT:

Northeast—April—July
Southeast—March—August
Midwest—April—July
Northwest—March—August
Southwest—March—August
Far North—April—July
Coastal—August-April

Soil: Rich loam produces flavorful, tender bulbs; pH: 6-7 (slight acid to neutral). Nutritional Requirements: calcium.

Planting: Add lime to soil before planting. Sow seeds ½ inch deep directly into garden or start seeds in indoor seedbeds in a cool room 4-6 weeks early. Space plants 1 foot apart.

Temperature: Kohlrabi is a Cool Weather plant. When grown during hot weather, it produces tough, fibrous bulbs. Can withstand very light frost, but should be planted during frost-free months.

Water: Plentiful water to encourage steady growth. Keep earth continuously moist.

Seasonal Protection: Mulch heavily to conserve moisture and protect against light frost.

Planting Seed to Harvest Time: 2 months.

Harvest: Cut below bulb when 2 inches wide.

LEEKS

WHEN TO PLANT:

Northeast—April-May
Southeast—February-April
Midwest—April-May
Northwest—March-May
Southwest—February-April
Far North—April-May
Coastal—February-April

Soil: Rich, well drained loam; pH: 6-8 (slight acid or alkaline to neutral). Nutritional Requirements: nitrogen.

Planting: Start seeds directly in garden, planting them ½ inch deep, or begin them 1 month early in indoor seedbeds. Space plants 6 inches apart.

Temperature: Leeks are Semihardy and a Cool Weather plant. They thrive best in mild weather and should be harvested before heavy frost.

Water: Abundance of moisture, especially in early stages of development.

Seasonal Protection: When plants are almost fully developed, gather earth up around stalks to where leaves begin. This will improve the flavor. Mulch.

Planting Seed to Harvest Time: 3-5 months.

Harvest: Pull up plant while still young and tender (3 months) or when fully mature (5 months).

LETTUCE

WHEN TO PLANT:
- Northeast—March & July
- Southeast—August-March
- Midwest—March & July
- Northwest—March & July
- Southwest—August-March
- Far North—April & July
- Coastal—August-March

Soil: Rich soil that retains moisture, yet has adequate drainage; pH: 7 (neutral). Nutritional Requirements: nitrogen, calcium, and phosphorus.

Planting: Seeds may be sown directly in garden during cool weather or started in indoor seedbeds in a cool room 3-5 weeks early. It is best to plant continuous seedings throughout planting period to provide a continuous supply.

Temperature: Lettuce is a Semihardy and Cool Weather plant. It should not be planted during the hot months (unless well shaded) as it thrives best in a cool, humid atmosphere. Some varieties are more heat-hardy than others and it is possible to grow lettuce during the summer under special conditions.

Water: Abundant at all times.

Seasonal Protection: Mulch heavily to conserve moisture. Provide shade by means of cloth or wooden covers set up to shade plants as hot weather approaches. If properly constructed, these shades may provide a means of growing lettuce during hot summer months. Lettuce will become bitter and bolt to seed if it receives too much light. It may also be planted in the shade of other vegetables.

Planting Seed to Harvest Time: 2-3 months.

Harvest: Pick leaves individually or cut head off at base. Roots will continue to send up new leaves.

MUSHROOMS
WHEN TO PLANT:

Mushrooms may be grown year round in a place that maintains the proper conditions of mild temperature and high humidity.

Soil: An underlayer of manure and hay covered with an inch of rich loam; pH: 7-8 (neutral-alkaline). Nutritional Requirements: nitrogen.

Planting: It is best to prepare a shed, cellar or cave in which to grow mushrooms, as a control of growing conditions is essential. Prepare a seedbed with a 6-inch layer of manure and hay mixed. Moisten and turn this mixture to aerate it until mixture has cooled to room temperature. Place mushroom spawn 1 inch deep throughout bed. Water regularly to keep moist and place in a room that has an even temperature of about 60 degrees. When mycelia spread throughout entire bed (in approximately 2-3 weeks) spread a 1 inch layer of rich loam on top of them. Mushrooms will develop 1½-2 months later.

Temperature: Mushrooms require just the right growing conditions, which include a steady 50-65 degree temperature, high humidity at all times, and an absence of light.

Water: Continuous.

Seasonal Protection: A thermometer and hygrometer to measure humidity is practically essential to be certain that proper growing conditions are maintained. To maintain the 50-65 degrees a heating or cooling system may be necessary, depending on outside climate. To maintain a high humidity in dry areas, a burlap cloth is useful, installed several inches above seedbed so that air circulates freely. This cloth should be kept wet at all times.

Planting Spawn to Harvest Time: 2-3 months.

Harvest: Cut at base of mushroom.

MUSTARD

WHEN TO PLANT:

    Northeast—April-May & August
    Southeast—February-April & August-September
    Midwest—April-May & August
    Northwest—March-May & August
    Southwest—February-April & August-October
    Far North—April-May & August
    Coastal—August-May

Soil: Porous loam; pH: 6-7 (slight acid to neutral). Nutritional Requirements: minimal.

Planting: Sow seeds ½ inch deep directly in garden or start in indoor seedbeds 2 weeks early. Space plants 6 inches apart. Successive plantings once a week will provide a continuous crop.

Temperature: Mustard is Hardy and a Cool Weather plant. It will survive light frosts.

Water: Continuous supply essential.

Seasonal Protection: Mulch to conserve water.

Planting Seed to Harvest Time: 1-1½ months.

Harvest: Pick leaves when young and tender.

OKRA

WHEN TO PLANT:

Northeast—May-June
Southeast—April-July
Midwest—April-June
Northwest—April-June
Southwest—April-July
Far North—May-June
Coastal—April-July

Soil: Moderately rich loam. If it is too rich it will cause plants to produce many leaves but few pods; pH: 6-8 (close to neutral). Nutritional Requirements: minimal.

Planting: Sow seeds ½ inch deep directly in garden or start seeds 3 weeks early in indoor seedbeds.

Temperature: Okra is Not Hardy and is a Warm Weather plant.

Water: Constant supply necessary.

Seasonal Protection: Heavy mulching important to conserve moisture.

Planting Seed to Harvest Time: 1½-2 months.

Harvest: Pick pods just after flowers drop off.

ONIONS [also scallions, green onions and shallots]
WHEN TO PLANT:

Northeast—March-May
Southeast—March-April & September
Midwest—March-May
Northwest—March-May
Southwest—September-March
Far North—March-May
Coastal—September-March

Soil: Rich, well drained sandy loam; pH: 5.5-6.5 (slightly acid). Nutritional Requirements: nitrogen, phosphorus, potassium, and trace minerals.

Planting: Sow seeds ½ inch deep or sets 1-2 inches deep directly in garden, or start seed in indoor seedbeds 8 weeks early. Space green onions 3 inches apart, drying onions 5-6 inches apart.

Temperature: Onions are Hardy and Frost Resistant. They do best in cool weather when forming green tops and hot weather when forming bulbs.

Water: Constant moisture important to keep bulbs from splitting open.

Seasonal Protection: Mulch heavily to conserve moisture.

Planting Seed to Harvest Time: Green: 1½-2 months; Dry: 3½-4½ months.

Planting Sets to Harvest Time: Green: 1 month or less; Dry: 2½-3 months.

Harvest: Pull green onions when bulb is ½-1 inch wide. Pull dry onions after tops have become dry and bulb is fully formed.

PARSNIPS

WHEN TO PLANT:

    Northeast—July-September
    Southeast—September-February
    Midwest—July-September
    Northwest—July-September
    Southwest—September-February
    Far North—June-August
    Coastal—September-February

Soil: Deep, moderately rich loam; pH: 6-8 (close to neutral). Nutritional Requirements: minimal.

Planting: Seeds germinate very slowly and should be soaked first. Plant directly in garden ½ inch deep or start in indoor seedbeds 4 weeks early. Space plants 3-6 inches apart.

Temperature: Parsnips are Very Hardy and a Cool Weather plant. Frost improves flavor.

Water: Plenty of water needed.

Seasonal Protection: Mulch heavily to conserve moisture.

Planting Seed to Harvest Time: 3-4 months.

Harvest: Pull after the first few frosts of the season.

PEANUTS

WHEN TO PLANT:

Northeast—April-May
Southeast—April
Midwest—April-May
Northwest—April-May
Southwest—April
Far North—Growing season too short
Coastal—April

Soil: Very porous, moderately rich soil; pH: 5-6 (low acid). Nutritional Requirements: calcium.

Planting: Plant raw peanuts or those sold for seeding, either in the shell or shelled 1½-3 inches deep, spacing plants 2 feet apart.

Temperature: Peanuts are Not Hardy and are a Hot Weather plant. They prefer a hot, sunny place in the garden.

Water: Continuous supply of water is needed up until 2 weeks before harvest.

Seasonal Protection: The peanut plant produces flowers which are pollinated and develop from the base of the petals small pod-like fruit. These pods bend over and penetrate the ground, developing into peanuts. Care should be taken to provide loose earth for these pods to enter. This may be done by placing a shovel-full of light humus around the base of each plant when it begins flowering. Mulch between plants to conserve moisture.

Planting Seed to Harvest Time: 4-5 months.

Harvest: When leaves begin to turn yellow, dig up plant and allow peanuts, plant and all, to dry by hanging in a dry place for a month or so.

PEAS

WHEN TO PLANT:

Northeast—March-May
Southeast—February-August
Midwest—March-May
Northwest—March-May
Southwest—February-August
Far North—April-June
Coastal—February-August

Soil: Light sandy loam hastens maturity of peas; pH: 6-6.5 (slightly acid). Nutritional Requirements: phosphorus, calcium and potash.

Planting: Sow seeds 2 inches deep directly into garden or start in indoor seedbeds 4-6 weeks early. Space plants 3-6 inches apart.

Temperature: Peas are Semihardy and are a Cool Weather plant: They will survive light frosts and will also thrive in warm weather. They prefer a cool, semi-shaded part of the garden, though it is the hot sunlight that causes them to burst into bloom. A cool spot will help to prolong the bearing period.

Water: Plenty of water is necessary, especially when pods are developing.

Seasonal Protection: Mulch to conserve moisture. Provide trellis, fence, or posts for peas to climb on.

Planting Seed to Harvest Time: 1½-2½ months.

Harvest: Pick edible-pod peas while still young and tender, or after peas have developed for shelling.

PEPPERS [green, banana, and hot]
WHEN TO PLANT:

    Northeast—May
    Southeast—May
    Midwest—May
    Northwest—May
    Southwest—May-June
    Far North—May
    Coastal—April-May

Soil: Rich and sandy with good drainage and plenty of humus; pH: 6-6.5 (slight acid). Nutritional Requirements: phosphorus and potash.

Planting: May be planted directly in garden, sowing seed ½ inch deep in warm soil, but it is usually best to start plants 6-8 weeks early in indoor seedbeds. Do not plant hot peppers near sweet peppers, as they may cross pollinate. Space plants 1-2 feet apart.

Temperature: Peppers are Not Hardy and are a Warm Weather plant. They prefer a mildly warm, humid climate.

Water: Continuous supply, especially in the early stages of growth.

Seasonal Protection: Mulch heavily to conserve moisture and protect from chilly winds at night.

Planting Seed to Harvest Time: 3-4 months.

Harvest: Cut stems when peppers feel heavy and solid.

POTATOES

WHEN TO PLANT:

Northeast—April
Southeast—February-March & August
Midwest—April
Northwest—March-April
Southwest—February-March & August
Far North—April-May
Coastal—February-March & August

Soil: Light loam with good drainage; pH: 4.5-6.5 (acid). Nutritional Requirements: do not use lime or wood ashes.

Planting: Use potatoes from a local grower, if possible, and be sure they are disease-free. Plant whole potato or cut into pieces about the size of a small lemon, with each piece containing 1 or 2 eyes (sprouts). Sow tubers 4 inches deep and space them 2 feet apart.

Temperature: Potatoes are Semihardy and a Cool Weather plant. They will survive light frost.

Water: Moderate watering. Do not let earth become soggy. Stop watering when leaves turn yellow.

Seasonal Protection: Mulch heavily to keep soil cool and protect developing tubers from sunlight. Gather earth up around developing plants, as tubers often develop above the soil surface and must be protected from light.

Planting Seed to Harvest Time: 3-4 months.

Harvest: Dig with a digging fork after flowers develop for young potatoes or allow to stay in ground until after the first few frosts. Allow to dry before storing.

POTATOES, SWEET

WHEN TO PLANT:
- Northeast—May
- Southeast—May-June
- Midwest—May
- Northwest—May
- Southwest—April-June
- Far North—May
- Coastal—April-May

Soil: Light, sandy, well drained loam; pH: 5.5-6.5 (acid). Nutritional Requirements: phosphorus and potash.

Planting: Select smooth, well shaped potatoes from last year's crop. Select a warm site such as the south side of a building or in hothouse. Place a four-inch layer of fresh manure on ground. Spread a three-inch layer of topsoil over manure. Lay potatoes on topsoil about 3 inches apart and cover with 2 inches of sandy loam. Potatoes will soon sprout "slips" which can be pulled and planted in garden when they are from 4-6 inches long. Sweet potatoes should be planted on raised rows 16 inches wide and 6 inches high.

Temperature: Sweet potatoes are Not Hardy and are a Warm Weather plant. They will not withstand frosts and grow best in a warm soil.

Water: Water continually until plants take root and vines begin growing. Gradually hold back on watering to encourage root rather than foliage growth.

Seasonal Protection: Mulch heavily. Gather earth up around plant base to protect developing tubers from sunlight.

Planting Slips to Harvest Time: 3-3½ months.

Harvest: Dig up with a digging fork the day after the first winter frost or after leaves turn yellow. Dry before storing.

RADISH

WHEN TO PLANT:

Northeast—March-May & August
Southeast—March-April & August-September
Midwest—March-May & August
Northwest—March-May & August-September
Southwest—August-March
Far North—March-June & August
Coastal—Year 'round

Soil: Rich loam with good drainage; pH: 6-8 (close to neutral). Nutritional Requirements: abundant nutrients near the surface.

Planting: Sow seeds directly in garden ½ inch deep and 1 to 3 inches apart, depending on variety. There are different varieties for spring (small red), summer (long white), and winter (long or round, white, red, or black). Continual planting throughout growing season is best.

Temperature: Radishes are Very Hardy and a Cool Weather crop. They will withstand frost. In long hot summer days, they bolt to seed quickly.

Water: Abundant to encourage rapid growth.

Seasonal Protection: Mulch to preserve soil nutrients and water often to keep ground cool.

Planting Seed to Harvest Time: 3 weeks to 2 months, depending on variety.

Harvest: Pull as soon as they reach full development, when small varieties are about 1 inch in diameter and large varieties are about 2-3 inches in diameter.

RHUBARB
WHEN TO PLANT:

May be planted in Fall and Spring in all areas that have a long cool growing season and a period of freezing so that the plants may lie dormant.

Soil: Well drained heavy loam; pH: 5.5-6.5 (low acid). Nutritional Requirements: plenty of everything, especially nitrogen.

Temperature: Rhubarb is Very Hardy and is a Cool Weather plant. It survives and actually welcomes long cold periods or periods of frost.

Planting: Plant roots or root crowns 3 inches deep and space 3 to 4 feet apart.

Water: Continuous supply throughout the year.

Seasonal Protection: Rhubarb is a perennial and will continue to produce for about 5-10 years. Spread a layer of mixed compost, manure and hay around the base of each plant every winter.

Planting Root to Harvest Time: 1 to 1½ years.

Harvest: Cut off large stems in spring or summer a year or so after planting. Yield: 15-20 stalks per year. Do not eat leaves, as they are very poisonous.

RICE

WHEN TO PLANT:

    Northeast—March-April
    Southeast—March-April
    Midwest—March-April
    Northwest—March-April
    Southwest—March-April
    Far North—Unsuitable climate
    Coastal—February-March

Soil: Rich bog or swamp-type condition; pH: 6-7 (slight acid to neutral). Nutritional Requirements: nitrogen.

Planting: Spread seed (with hulls) over an area that can be flooded during the growing season.

Temperature: Needs a long, hot growing season.

Water and Seasonal Protection: As plants begin growing, gradually flood area with water, preferably water which is circulating or flowing slowly. When plants reach maturity, water should be 6 to 9 inches deep.

Planting Seed to Harvest Time: 4-5 months.

Harvest: Shake seeds from stalk. Dry them in the sun for a week. Heat in a pot over a low flame until hulls are very dry, stirring continually. Place in a sack and pound with a stick to separate kernels from hulls. Separate grain from chaff by tossing into the air over a cloth on a windy day.

RICE, WILD
WHEN TO PLANT:

Northeast—October-November & March-April
Southeast—November-December & Feb.-March
Midwest—October-November & March
Northwest—October-November & March
Southwest—November-December & Feb.-March
Far North—October
Coastal—November-March

Soil: Swamp or surrounding a pond; pH: 6-7 (slight acid to neutral). Nutritional Requirements: nitrogen.

Planting: Soak seed for a day. Spread seed (with hulls) in an area where water is slow moving and depth at a fairly constant level.

Temperature: Wild Rice is Hardy and will survive cold, frosty winters, coming back up again in the spring.

Water: Water level should be a half foot to 3 feet deep and not fluctuating.

Seasonal Protection: Wild Rice is a perennial and will take care of itself, continuing to reseed itself after the first planting.

Planting Seed to Harvest Time: 5-7 months.

Harvest: Shake or beat seeds from stalk. Dry them in the sun for a week. Heat them in a pot over a low flame until hulls are very dry, stirring continually. Place in a sack and pound with a stick to separate kernels from hulls. Separate grain from chaff by tossing into the air over a cloth on a windy day.

RUTABAGA

WHEN TO PLANT:

Northeast—March-April & July
Southeast—August-March
Midwest—March-April & July
Northwest—March-April & July-August
Southwest—August-March
Far North—April & July
Coastal—August-March

Soil: Rich loam, well drained; pH: 5.5-7 (low acid to neutral). Nutritional Requirements: nitrogen.

Planting: Sow seeds ½ inch to 1 inch deep directly into garden or start in indoor seedbeds 4-6 weeks early. Space plants 1 foot apart.

Temperature: Rutabagas are Very Hardy and are a Cool Weather plant. They will not last very long in frozen ground.

Water: Fairly constant, especially in dry weather.

Seasonal Protection: Mulch to conserve moisture.

Planting Seed to Harvest Time: 3 months.

Harvest: Pull up roots when 3 inches wide or just after the first heavy frost.

SESAME SEEDS

WHEN TO PLANT:
May in all areas.

Soil: Moderately rich loam; pH: 6-8 (close to neutral). Nutritional Requirements: nitrogen.

Planting: May be sown directly in garden in areas with a long growing season. In other areas, start seeds in warm indoor seedbeds 6-8 weeks early. Sow seed ½ inch deep and 6 inches apart.

Temperature: Sesame Seeds are Not Hardy and are a Warm Weather plant. They need a long, hot growing season.

Water: Moderate.

Seasonal Protection: Protect from cold breezes, chilly nights, and heavy rain.

Planting Seed to Harvest Time: 3-4 months of hot weather.

Harvest: Pick seed pods before the first frost.

SOYBEANS

WHEN TO PLANT:

Northeast—May
Southeast—April-June
Midwest—May-June
Northwest—May
Southwest—April-June
Far North—May
Coastal—April-June

Soil: Rich loam, well drained; pH: 6-7 (low acid to neutral). Nutritional Requirement: lime, phosphorus, potash.

Planting: Sow seeds 1 inch deep directly in garden or start in indoor seedbeds 4-6 weeks early. Space plants 6-12 inches apart.

Temperature: Soybeans are Not Hardy and are a Warm Weather plant, though they are more hardy than other varieties of bean.

Water: Constant supply essential.

Seasonal Protection: Mulch heavily to conserve moisture.

Planting Seed to Harvest Time: 2½-3½ months, depending on variety.

Harvest: For green soybeans: Pick when nearly mature, before pods turn yellow, and boil for one minute to make shell removal easy. For dry soybeans: Allow beans to dry on plant but pick before pods and stem become brittle.

Planting Soybeans for Green Manure: Turn entire plants under the ground when pods begin forming.

SPINACH

WHEN TO PLANT:

Northeast—March-April & August-September
Southeast—August-March
Midwest—March-April & August-September
Northwest—March-April & August-September
Southwest—August-March
Far North—April & July-September
Coastal—August-April

Soil: Rich, deep loam; pH: 6-6.5 (this range very important) (low acid). Nutritional Requirements: nitrogen.

Planting: Sow seed ½-1 inch deep directly into garden or start in indoor seedbeds in a cool room 2 weeks early. Space plants 6-12 inches apart. It is good to plant continually, every 1-2 weeks during planting season, to provide a continuous crop.

Temperature: Spinach is Hardy and a Cool Weather crop. It tends to bolt to seed in hot weather.

Water: Continuous supply. It should be irrigated during dry spells.

Seasonal Protection: Mulch (NOT with a high-acid mulch such as oak leaves or redwood sawdust) during warm season to keep plants cool, and during cold season to protect against heavy frosts.

Planting Seed to Harvest Time: 1½-2 months.

Harvest: Cut leaves individually when young and tender, or pick entire plant when developed with several large leaves.

SPINACH, NEW ZEALAND

WHEN TO PLANT:

Northeast—April-July
Southeast—March-August
Midwest—April-July
Northwest—April-July
Southwest—March-August
Far North—April-July
Coastal—April-August

Soil: Deep rich loam; pH: 6-7 (low acid to neutral). Nutritional Requirements: nitrogen and lime.

Planting: Soak seed in warm water overnight and sow directly into garden 1½ inches deep or start in indoor seedbeds 4-6 weeks early. Space plants 1-2 feet apart.

Temperature: New Zealand Spinach is Not Hardy and is a Warm Weather plant. It survives in hot weather much better than regular spinach, to which it is not even closely related, but does not like extremely hot weather, nor is it frost hardy.

Water: Continuous supply at all times.

Seasonal Protection: Protect from extreme heat with shading and mulch to conserve water.

Planting Seed to Harvest Time: 2½ months.

Harvest: Pick leaves when young and tender. Plant will continue to produce until frost arrives.

SQUASH, SUMMER [crookneck, coccozelle, scallop, zucchini]

WHEN TO PLANT:

Northeast—May-July
Southeast—April-August
Midwest—May-July
Northwest—May-July
Southwest—April-August
Far North—May-June
Coastal—April-August

Soil: Rich sandy loam; pH: 6-7 (low acid to neutral). Nutritional Requirements: nitrogen.

Planting: Sow seed directly into garden 1 inch deep or start in indoor seedbed 3-4 weeks early. Space plants 3-4 feet apart.

Temperature: Summer Squash are Not Hardy and are Warm Weather plants. They enjoy hot daytime temperatures.

Water: Continuous supply.

Seasonal Protection: Mulch to conserve moisture and to prevent squash from touching damp earth.

Planting Seed to Harvest Time: 2-2½ months.

Harvest: Pick when young and tender.

SQUASH, WINTER [acorn, banana, boston, butternut, buttercup, cushaw, hubbard and pumpkin]

WHEN TO PLANT:

Northeast—May-June
Southeast—April-July
Midwest—May-June
Northwest—May-June
Southwest—April-July
Far North—May-June
Coastal—April-July

Soil: Rich sandy loam; pH: 6-7 (low acid to neutral). Nutritional Requirements: nitrogen.

Planting: Soak seed overnight in warm water. Sow seed directly in garden 1 inch deep or start in indoor seedbeds 4 weeks early. Space the plants 6 feet apart, as they need plenty of room to spread out.

Temperature: Winter Squash are Not Hardy and are Warm Weather plants. They do well in hot weather.

Water: Continuous supply.

Seasonal Protection: Mulch to conserve moisture and to prevent squash from touching the damp earth. Provide plenty of room for vines to grow, or fence, trellis or ladder for them to climb.

Planting Seed to Harvest Time: 3-4 months.

Harvest: Cut stems just before first frost and allow to dry for a few days in the sun or in a frost-free place. Wash and dry, handling carefully and store in a cool place.

STRAWBERRIES

WHEN TO PLANT:

Northeast—March
Southeast—February-March
Midwest—March
Northwest—February-March
Southwest—February-March
Far North—March-April
Coastal—December-February

Soil: Well drained, rich, loose, sandy loam; pH: 5.5-6.5 (low acid). Nutritional Requirements: plenty of organic matter.

Planting: Plant cuttings (runners) in early spring while they are dormant, or start whole plants and keep ground moist while roots are developing and spreading. Allow plenty of space, as strawberries spread rapidly. Do not plant new strawberries in an old strawberry bed.

Temperature: Strawberries are Hardy once they are established, but prefer mild, warm weather.

Water: Large quantities, especially while berries are forming. Moderate quantities after berries have been harvested.

Seasonal Protection: Strawberries are perennial and each plant will continue to produce for many years. Mulch to protect fruit from touching damp ground and to preserve moisture. The first summer, all blossoms should be picked so that the plant will put total energy into developing its root system. In fall, prune runners so that bed won't be crowded. After the harvest fertilize by placing a side dressing of compost around each plant. In the fall, spread a layer of mulch over the top of the plants and remove it in the spring.

Planting Cutting to Harvest Time: 6 months to 1½ years.

Harvest: Pick berries, removing stem, when red and ripe.

SUNFLOWER
WHEN TO PLANT:

Northeast—March-June
Southeast—February
Midwest—March-June
Northwest—March-June
Southwest—February
Far North—March-May
Coastal—February

Soil: Rich deep loam, somewhat sandy; pH: 6.5-7 (low acid to neutral). Nutritional Requirements: nitrogen and phosphorus.

Planting: Sow seed ½ inch deep directly into garden and space plants 1-2 feet apart. It will hasten germination to soak seeds in warm water overnight.

Temperature: Sunflower is Semihardy, but is a Warm Weather plant. Young plants will survive light frost and mature plants will survive heavy frosts.

Water: Moderate.

Seasonal Protection: Provide supports, such as poles for varieties that are very tall, and for all plants when seedheads develop, to support the heavy heads. Mulch to conserve moisture.

Planting Seed to Harvest Time: 3-3½ months.

Harvest: Cut off head when brown and dry or when still slightly green. If picked when green, allow to dry in a warm room.

TOMATOES
WHEN TO PLANT:
Northeast—May-June
Southeast—April-June
Midwest—May-June
Northwest—May-June
Southwest—April-June
Far North—May
Coastal—April-June

Soil: Moderately rich, slightly sandy loam; pH: 6-7 (low acid to neutral). Nutritional Requirements: phosphorus and potassium (do not add too much nitrogen).

Planting: Sow seeds directly into garden ½-1 inch deep or start in indoor seedbeds 7-8 weeks early. It is often good to plant several varieties for a continuous supply. Space plants 2-3½ feet apart.

Temperature: Tomatoes are Not Hardy and are a Warm Weather plant, favoring hot days and warm nights in a very sunny location. Some plants will survive light frosts.

Water: Water thoroughly, but not too heavily after fruit begins forming. Hold off water when you want them to start ripening.

Seasonal Protection: Mulch heavily. Protect from hot, dry wind and from intense heat when setting out transplants by using an inverted wooden crate, barrel or basket with a few open spaces. Place this over the seedling. It may also be left in place after vine begins developing, as a form of support on which the vines may grow. If you do not use this method, provide a trellis, tripod, fence or stake on which they may be tied. Lack of potassium will cause fruit to ripen unevenly.

Planting Seed to Harvest Time: 2-4 months.

Harvest: Pick when fully ripened and before the first heavy frost.

TURNIPS

WHEN TO PLANT:

Northeast—April & August
Southeast—August-April
Midwest—April & August
Northwest—April & August
Southwest—August-April
Far North—April & July
Coastal—August-April

Soil: Rich, well drained loam; pH: 6-8 (close to neutral). Nutritional Requirements: nitrogen and phosphorus.

Planting: Sow seeds thinly ½ inch deep directly into garden or start in indoor seedbeds in a cool room 2 weeks early. Space the plants 6 inches apart.

Temperature: Turnips are Hardy and are a Cool Weather plant. May bolt to seed in hot weather.

Water: Moderately moist.

Seasonal Protection: Mulch to conserve water.

Planting Seed to Harvest Time: 1-2 months.

Harvest: Pull roots or pick greens to eat raw when young and tender.

WATERMELONS
WHEN TO PLANT:

Northeast—May
Southeast—April-May
Midwest—May
Northwest—May
Southwest—April-June
Far North—May
Coastal—April-June

Soil: Loose, sandy loam; pH: 5.5-6.5 (low acid). Nutritional Requirements: plenty of everything.

Planting: Plant seeds directly in the garden ½ inch deep or start seeds in warm indoor seedbeds 6-7 weeks early. Allow plenty of space, as they spread rapidly. Plant groups of 2 or 3 spaced 6 feet apart.

Temperature: Watermelons are Not Hardy and are a Warm Weather plant. They need full sun while growing.

Water: Abundant supply at all times.

Seasonal Protection: Mulch to conserve moisture, to control weeds, and to protect plants from touching damp earth.

Planting Seed to Harvest Time: 2½-4 months, depending on variety.

Harvest: Pick when melon makes a low dull sound when tapped with knuckles. An unripe melon has a high, tinny sound.

WHEAT

WHEN TO PLANT and VARIETY:

Northeast—April—Hard Red Winter Wheat
Southeast—April-May—Soft Red Winter
Midwest—April—Hard Red Winter
Northwest—April—Hard Red Winter
Southwest—April-May—White Wheat
Far North—April—Hard Red Spring and Durum
Coastal—April—Soft Red Winter and White

Soil: Well tilled, weed-free, rich loam, somewhat heavy; pH: 6-7 (low acid to neutral). Nutritional Requirements: lime, phosphorus, and nitrogen.

Planting: Choose a variety of wheat that is hardy to your local conditions. Sow seed thickly in rows and cover with 1-2 inches of earth.

Temperature: Wheat is Hardy and a Cool, Dry Weather plant. It does not like extremely humid weather, though some varieties are more tolerant of various growing conditions.

Water: Moderate while seeds are germinating, and minimal after plants begin to grow.

Seasonal Protection: May be mulched between rows.

Planting Seed to Harvest Time: 3½ months.

Harvest: Cut wheat near the base of the stalk. Let grain dry on the stalk for a week or so. Remove seedheads and place in a pot over a low flame until the hulls are very dry, stirring continually. Place in a sack and pound with a stick to separate kernels from hulls. To separate grain from chaff, toss into the air over a cloth on a breezy day.

WATERING THE GARDEN

All living things on the Earth need water to grow. Each has its own requirements and reactions to the amount of water it receives and the time at which it receives it. Water in itself is simply hydrogen and oxygen. Its importance lies in its ability to carry vital nutrients to living cells.

In the garden, it is water that carries the nutrients that plants need as well as the moisture that creates a healthy atmosphere for plants to thrive. Areas that have irregular rainfall will need to have an irrigation system that can be as simple or as sophisticated as energy, terrain and finances allow. Here are some important considerations when watering a garden.

1. In Clay or other heavy soils, be careful not to over-water. As soil drainage in heavy soils is poor, a water-logged soil may drown plants.

2. In Sandy or other light soils, constant or frequent watering may be essential, as water will leach through rapidly.

3. Water in the morning! This is especially important in areas where night-time temperatures drop below 40 degrees. In the early morning, the earth is colder than the water. By watering, you warm the earth which encourages growth. As the moisture evaporates during the warm part of the day, it provides the necessary humidity for the plants. In the evening, the earth is warmer than the water and watering at this time chills the earth. Watering at night also encourages fungus-type growth on and around plants.

4. When using reservoir water, it is best to use water as close to the surface of the reservoir as possible, as this water is the warmest.

5. Water deeply. Shallow watering, which is brought about by not giving plants enough water, causes roots to grow upward in search of water. Proper, thorough watering allows roots to grow deeply, which is important for plants to get the proper nutrients from the soil.

6. Sprinkle lightly just after planting seeds to avoid seed's washing out or falling into trenches between rows.

7. Do not water while the sun is shining intensely on the garden, as each drop of water sitting on a plant leaf acts as a magnifying glass and plant leaves may burn.

MULCHING

Mulch is a protective layer of loose material which shields the base of the plant as well as the surrounding earth from a variety of natural elements.

Mulching is done by spreading a 2- to 6-inch layer of material between rows of plants and around the base of each plant. Materials commonly used for mulch are:

Buckwheat Straw and Hulls: Excellent. Adds humus to soil, does not pack down because it is loose and light in weight. Does not alter soil pH.

Compost: Excellent. Adds valuable nutrients to soil. A variety of raw wastes may be used, such as citrus peels, carrot tops, corn cobs, eggshells, etc. Will balance soil pH.

Hay and Straw: Fair. Is a good texture and often is readily available. It may cause a problem by adding seeds to the garden, bringing weed problems. It also tends to mat down under heavy rains, preventing aeration of the soil. Adds humus and nitrogen to the soil and does not alter soil pH.

Hulls and Shells: Excellent. Will not decompose rapidly. Adds nitrogen and potassium to the soil. Remains loose, yet heavy enough to stay in place. Some shells increase acidity.

Last Season's Dried Husks and Leaves: Good. Adds humus to the soil. Corn husks and stalks, pea vines, squash leaves, and a variety of other vines, stems and leaves may be used. Legume refuse adds nitrogen to the soil. Improves soil pH.

Leaves: Excellent. Are often readily available in large quantities. Add humus, as well as calcium, nitrogen, phosphorus, potassium and magnesium to the soil, and are good as long as they do not pack down in wet weather. All tree leaves are acid and should be leached before using. This may be done by collecting them from pond or creeks after winter rains have washed them. A simple way to leach them yourself is to stretch a 2-foot high piece of chicken mesh across a stream or on a hillside downhill from a group of deciduous trees. Leaves will fall and be washed down by winter rains, leaching acid from them. In a very alkaline soil, these leaves may be used unleached, as their acidity will help to neutralize the soil. They may also be used unleached if limestone is added to the garden along with the leaf mulch. Keep a check on soil pH.

Peat Moss: Good. This is the partially decomposed remains of plants, formed in swamplands and other bodies of still water. It is light and very absorbent. It adds some humus to the soil, but is highly acid and should be used in alkaline soils, or where limestone is added, or for acid-loving crops.

Pond Plants: Excellent. A variety of pond life (such as coontail, mentioned earlier) may be used as a mulch when it is readily available in large quantities. These plants add humus and nutrients to the soil and will help to balance soil pH.

Hardwood Sawdust: Good. Should not be used in great quantities and should be mixed with a looser mulch material.

Seaweed: Good. Should be washed first, as it contains much salt. Adds nitrogen, potassium and phosphorus to the soil. Increases alkalinity.

Stones: Good. Spread loosely enough to allow for soil aeration. Stones add minerals and help warm garden soil. Stones containing limestone will increase soil alkalinity.

Wood Chips (Hardwood): Good-Fair. A good texture to work with. Often available at lumber mills. Should be leached first, as some are high in acid and will increase soil acidity. This may also be counteracted by adding limestone when mulching.

Nearly any material available to you in quantity may be used as a mulch. Mills and food processing plants often have various refuse materials they would be glad to give away. Avoid any materials that have been sprayed with chemicals of any sort, as they may cause damage to your crops. Materials used must be loose enough to allow for aeration of the soil surface.

Mulching has many purposes:

Prevents soil from drying out.

Protects against light frosts.

Reduces irrigation requirements.

Controls weeds.

Lessens damage done by small burrowing rodents.

Helps ground retain coolness of the night during the day and warmth of the day during the night.
Adds humus to the soil.
Protects low growing fruit and vegetables from touching damp earth.
Winter protection for perennials.
Improves soil texture and fertility.
Protects against sudden changes in temperature.
Prevents soil from packing down and becoming tight.
Protects against extreme heat.
Protects against chilling wind.
Lessens erosion in the garden.
Helps control insects.

FERTILIZING

After the garden has become established, the store of minerals in the soil often becomes depleted unless new ones are added from time to time. It is important to use only organic fertilizing methods, as chemical fertilizers kill valuable soil bacteria that are essential for healthy plant growth. These bacteria help the plants to assimilate the available nutrients in the soil.

The arrival of insects and disease in the garden is often caused by mineral deficiencies in the soil. Each plant has its own particular nutritional requirements which must be met if the plant is to remain healthy and productive. The following is a list of minerals and their organic sources:

Nitrogen:

compost	legume cover crop	seaweed
bloodmeal	vegetable meal	manure
bonemeal	grass cuttings	leaves
fish wastes	dry legume stalks	hair
alfalfa hay	feathers	

Phosphorus:

| bloodmeal | fish wastes | seaweed |

bonemeal   manure   wood ashes
compost   phosphate rock

Potassium:

grass cuttings   potash rock   vegetable ashes
hulls   seaweed   vegetable peels
manure   shells   wood ashes

Magnesium:

compost   leaves   phosphate rock
dolomite   limestone

Calcium:

compost   leaves   phosphate rock
dolomite   limestone   wood ashes
eggshells   milk products

Manganese:

carrot tops   compost   hardwood tree leaves

Sulphur:

compost

Iron:

compost   seaweed   weeds
phosphate rock

Zinc:

alfalfa   oak leaves   sawdust
compost   phosphate rock   vetch

Boron:

clover   oak leaves   sawdust
compost   phosphate rock   vetch
melon leaves

Trace Minerals:

alfalfa   hardwood sawdust   phosphate rock
compost   hardwood tree leaves seaweed
corn stalks   legumes   vetch

Also see Improving Soil Fertility and Fertilizers.

255

SUPERLATIVE COMPOST DEVELOPED BY OREMS ORGANIC GARDENS

The use of magnesium sulphate (epsom salts) in the compost pile or directly in the soil is very beneficial in growing organic food. The food grown in soil that contains magnesium sulphate is very beneficial to your health. It greatly improves soil texture, as well as mineral content.

The use of skim milk or powdered skim milk in the compost pile or directly in the garden will greatly enhance the decomposition of the compost and the finished product will be unbelievably rich in valuable nutrients and beneficial bacteria.

Coontail (Hornwort) makes the best compost and soil builder known. It is easily grown in ponds or reservoirs. It can be removed from the ponds by a drag-rake hooked to a cable and pulled by a winch if it is cultivated in large quantities. To make the coontail most beneficial, the pond or reservoir should be stocked with a zooplankton such as Daphnia (water flea). When the coontail is removed from the pond, it may be put directly into the soil. It decomposes very rapidly into a rich compost. It may also be added to the compost pile and will aid in the rapid decomposition of the wastes.

A combination of the above will make the best compost known and, by using it properly, you will have unbelievable results with your organic garden. In a research project, ordinary Kentucky Wonder Beans grew leaves 17 inches wide and bean pods 12+ inches long. Another experiment grew 200 pounds of carrots from 20 square feet of growing space and 800 pounds of sweet potatoes from 52 square feet of space.

DEFICIENCY SYMPTOMS

Symptoms of Disease	Possible Deficiency or Cause
PLANT:	
growth slow or retarded	nitrogen, phosphorus, sulphur, potassium, boron.
wilts or droops	potassium, soil too alkaline, too much heat, roots cramped, insects.
infested with insects	calcium, not enough compost.
low resistance to disease	potassium, trace minerals.
seeds sterile	phosphorus, potassium.
abundant foliage with few fruit	over fertilization with too much nitrogen.
ROOTS:	
underdeveloped roots	calcium, soil too alkaline, lack of water.
tubers fail to develop	potassium.
STEMS:	
twisted, hardened stalks	nitrogen.
stem turns purple	nitrogen.
stalk does not develop	magnesium, phosphorus.
stem tips become spotted with red blotches	zinc.
stem turns soft and begins to decay	potassium, too much water, improper soil drainage.
stems and stalk thin and spindly	phosphorus, not enough light.
LEAVES:	
curled and misshapen	calcium, aphids.

Symptoms of Disease	Possible Deficiency or Cause
LEAVES: (cont'd)	
yellow stripes	calcium, insects.
edges ragged with brown blotches	calcium, insects.
yellow with gray blotches	manganese, insects, too much sun.
leaf tips curl downwards	sulphur, boron, insects.
pale green at base of plant	nitrogen, boron, insects, overwatering.
yellow at top of plant	nitrogen, iron, soil too alkaline or too acid.
yellow at base of plant	nitrogen, magnesium, soil too alkaline or too acid, extreme temperature changes, humidity too low, improper aeration, uneven watering.
red or purple	nitrogen
fall off prematurely	nitrogen, magnesium, zinc, overwatering, soil too alkaline, sudden temperature changes.
brown	magnesium, potassium, manganese.
dull or pale	phosphorus.
blue	phosphorus, potassium.
dull gray	potassium.
bronzed with purple edges	phosphorus, zinc.
bronzed with brown spots	phosphorus.
yellow splotches	potassium, zinc, insects, irrigation water too cold.
brown with curled edges and leaf tips	potassium, boron, insects, frost, too much fertilizer,

Symptoms of Disease	Possible Deficiency or Cause
LEAVES: (cont'd)	
	too much heat, low humidity, uneven watering, overhead watering during the day when sun is intense.
leaf edges irregular	calcium.
leaves turn white	zinc, insects, irrigation water too cold.
leaves become stiff	zinc.
wrinkle and fall off	zinc, boron.
swell and blister	too much water, too much fertilizer.
poor coloring	iron, magnesium, phosphorus.
holes in leaves	insects.
yellow with grainy undersides	red-spider mites.
curled with red and yellow blotches	mosaic (a virus disease)
leaves unusually dark	too much fertilizer.
FLOWERS:	
buds underdeveloped	nitrogen.
few flowers develop	potassium, boron, lack of water, lack of sun, too much fertilizer.
many flowers, few fruit	potassium, lack of pollination.
buds drop off prematurely	temperature too hot, humidity too low, sudden temperature extremes, heavy winds, fertilized too late in season.

Symptoms of Disease	Possible Deficiency or Cause
FRUIT:	
fails to develop	nitrogen, boron, magnesium, phosphorus, potassium, lack of pollination.
small and of poor quality	nitrogen, magnesium, potash.
soft and rots quickly	phosphorus, potassium.
bitter	phosphorus, potassium.
low yield	phosphorus, potassium, boron.
ripens unevenly	potassium.

FERTILIZERS

COMPOST

Compost is the main and most important fertilizer for an organic gardener. It completes the balanced cycle of life, returning organic wastes to the soil so that nothing is really wasted.

Compost contains all minerals necessary for healthy plant growth and presents them to the plants in an easy-to-assimilate form. It balances the soil pH, which insures the availability of the various minerals to the plants.

Compost improves soil texture, allowing proper drainage and aeration, and adds valuable humus to the soil. It is usually applied to the garden during the tilling or as a "side dressing" for already established plants.

MANURE

Manure is also part of the balanced cycle of life on earth. It contains a large quantity of nitrogen, as well as phosphorus and potassium, the three main essentials for healthy plant growth.

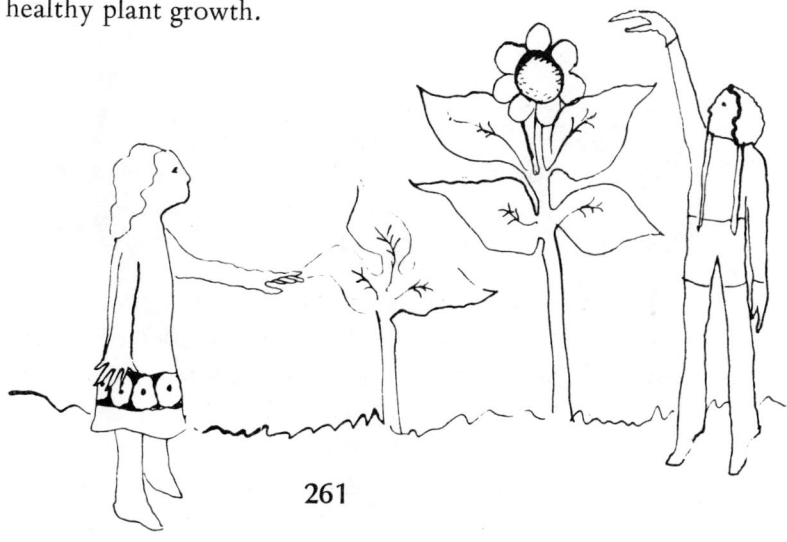

Manure is available in the form of:

green manure (plant wastes)
chicken   pig
cow   horse
sheep   human

Manure from the barnyard is often collected along with bedding material, such as straw, which further adds to its soil improving value.

Fresh manure should never be added directly to a plant or to a garden site about to be planted. It is so potent that it may cause injury to delicate plant roots. It should, instead, be added to the compost pile or dug into a garden site to be planted the following season, or added in the form of a weak solution. Never use manure on new seedlings, as this encourages rapid top growth and does not allow roots to develop fully. Do not add manure during mid-summer or fall, as this will encourage new growth that will not be hardened against the on-coming winter. Too much manure will cause abundant foliage growth but little fruit development, so use it carefully. Use it very sparingly or not at all on root crops.

COVER CROPS

Cover crops are often planted to prevent soil erosion on open ground, as a form of "green manure" to add minerals to the soil and to improve texture, or as a living mulch to protect crops from frost.

The crop, often a legume, is planted in the late summer or fall a few weeks before the harvest of the vegetable grown in the plot it is to cover. The cover crop is allowed to grow until the following spring, at which time it is tilled and turned into the soil.

The following cover crops add valuable nitrogen to the soil while growing and valuable humus when they are turned into the earth:

alfalfa	lupine
all legumes	millet
beans	peas
clover	soybeans
cow peas	vetch
lentils	

Alder trees, though not generally turned into the earth in spring, add nitrogen to the surrounding soil while growing and their leaves provide excellent humus material.

Cover crops which "choke out weeds" are:

clover	millet
crown vetch	St. Augustine grass

Cover crops which prevent erosion and add humus when they are turned into the soil are:

alfalfa	legumes
buckwheat	lupine
chickweed	lyme grass
clover	millet
coltsfoot	vetch
heather	wild rye

Cover crops which act as a "living mulch" are:

alfalfa	clover	millet

All legumes should be inoculated with the proper micro-organism inoculant available at seed stores. Follow instructions as mentioned on the package of inoculant.

Fertilizer (organic!) may be added to an already established garden in the following ways:

1. Side Dressing: Place a shovelful of compost humus, well rotted manure, and or mineral rock around the base of each plant or along the row of plants. Compost may be used in any quantity, but go lightly with the manure and mineral rock. The best way is to add manure and mineral rock to compost pile. This side dressing is left on the surface of the earth (under the mulch), dug into the soil a depth of a few inches, or placed in a trench dug a few inches below the surface.

2. As a "Tea": Dissolve compost, dried manure, or mineral rock in water, making a rather mild solution when using manure or mineral rock. "Manure Tea" should be about the color of tea. Irrigate area thoroughly with plain water. Apply weak solution to the earth, taking care not to get any on plant leaves or stem. Water again with plain water. Do not use "manure tea" more than once every 2 weeks or so.

3. Minerals are added to raw compost and allowed to "mature" along with the compost. This compost is later added as a side dressing.

4. The use of various mulches which contain nutrients: Watering carries nutrients down into soil and to the plant roots.

5. Minerals or dry manure may be added to mulch before the mulch is applied.

SEASONAL PROTECTION

A garden is not just a springtime venture. Constant

care and attention are necessary. As growing seasons vary from place to place, it is impossible to give a standard for everyone to follow. Through experience you will learn the nature of your own particular garden and what can be done each season to keep it productive and beautiful.

It is interesting to keep a garden notebook and record each new planting, its progress, cosmic as well as earthly information, drawings, and feelings about your garden.

WINTER

1. Mulch heavily around winter vegetables to protect them from frosts.
2. Dig drainage ditches in heavy-rain areas to drain water from low spots in the garden.
3. Begin planning the spring garden, and send away for seed if ordering through mail-order nurseries.
4. In severely cold areas, cover lettuce and herbs with a coldframe.
5. Collect fallen leaves for mulch or compost.
6. Collect vegetable plant remains for mulch or compost.
7. Remember to rotate compost for aeration at least once a month.
8. Protect compost heap from heavy rains and snow, as water will leach out valuable nutrients from compost.
9. Spread a 1-foot layer of mulch over already-established asparagus, artichoke, rhubarb, and strawberry plants.

10. Plant new perennials, such as artichoke, asparagus, rhubarb, and strawberry plants in garden, as they are dormant in the winter.

11. Test soil for nutrients and pH and add necessary elements just before spring in rainy areas, and in midwinter in dry areas.

12. In late winter, start in indoor seedbeds: cabbage, cauliflower, celery, Brussels sprouts, eggplant, green peppers, onions, and tomatoes.

SPRING

1. Add compost humus to garden site.
2. Add sand to garden site if soil is nonporous.
3. When frost danger is past, transplant seedlings started indoors.
4. Before frost danger is past, but after ground has warmed, these may be planted directly in garden: beets, broccoli, Brussels sprouts, cabbage, carrots, chard, collards, endive, garlic, Jerusalem artichokes, kale, kohlrabi, leeks, lettuce, mustard, onions, peas, potatoes, radishes, rice, rutabaga, spinach, turnips, wheat.
5. After frost danger is past, these may be planted: beans, celery, corn, cucumbers, eggplant, herbs, melons, New Zealand spinach, okra, peanuts, peppers, radishes, sesame seeds, soybeans, summer and winter squash, sunflowers, tomatoes, and watermelons.
6. Remember to rotate compost heap at least once a month to aerate it.

7. Plant every week or so throughout spring for a continuous supply: beets, carrots, lettuce, onions, peas, radishes.

8. Mulch carefully and not too heavily so as not to drown seedlings in spring rains or cause seeds to decay in cold, damp earth.

9. Sow a cover crop in the winter garden site.

10. Remove mulch cover from perennials and spread between rows of plants, but not too close to plant base, as the earth may become too damp from the spring rains.

SUMMER

1. Protect cool-weather plants from scorching sun with burlap or wooden shades. Be sure they get plenty of water.

2. Mulch heavily to conserve moisture and control weeds.

3. Continue planting every week for a continuous supply: beets, carrots, lettuce, onions, radishes.

4. Add a side dressing of compost around each plant.

5. In late summer, water tomatoes sparsely, as lack of water will induce ripening.

6. In midsummer, begin planting these for a fall garden: beets, broccoli, Brussels sprouts, cabbage, carrots, cauliflower, chard, endive, green onions, kale, kohlrabi, lettuce, mustard, radishes, spinach, and turnips. Plant these every week or so throughout end of summer.

7. Weed garden.

8. Irrigate carefully to keep ground continuously moist.

9. Do not fertilize perennials after midsummer, as this will cause new green growth that may easily be damaged when winter arrives.

FALL

1. In early fall, plant a cover crop on unused garden area, to be turned under in the spring.

2. After harvest of summer vegetables, add compost and other organic fertilizers to the area. This area may then be used for fall crops.

3. In very early fall, make one last planting of: beets, broccoli, Brussels sprouts, cabbage, carrots, chard, collards, endive, garlic, green onions, kale, kohlrabi, leeks, lettuce, mustard, radishes, spinach, and turnips.

4. In a cold frame plant: celery, lettuce, and parsley.

5. Harvest before the first heavy frost: beans, cucumbers, eggplant, herbs, melons, New Zealand spinach, okra, peanuts, peppers, sesame seeds, soybeans, squash, sunflowers, sweet potatoes, and tomatoes.

6. If frost comes before harvest time, pull up entire plants and hang upside down in a dry room.

7. The following plants will survive light fall frosts: chard, collards, corn, kale, lettuce, peas, radishes, root crops, spinach, other greens.

8. The following plants will survive moderately heavy frost: broccoli, cabbage, chard, collards, endive, kale, lettuce, mustard, radishes, root crops, and spinach.

9. The following plants are actually improved by frost: apples, broccoli, cabbage, collards, endive, kale, parsnips, persimmons, rhubarb.

10. Perennials which will return the following spring, sending forth new growth are: artichokes, asparagus, chard, endive, horse radish, Jerusalem artichokes, kale, parsley, rhubarb. These plants should be covered several inches deep with mulch to protect them over the winter.

11. Mulch around all winter crops to protect against harsh freezes.

12. Clean up all plant refuse and place in compost pile.

13. Add fallen leaves to compost pile or use as mulch.

14. Put away all garden tools to protect against rain.

15. Empty all exposed water lines to garden, except in areas of mild climate or areas that will need winter irrigation. This will prevent pipes from freezing and cracking.

16. Plant flower bulbs.

17. Add fresh manure to compost heap so that it will be ready for next spring.

18. Do not fertilize trees or other perennials, as this will encourage new growth that may be easily damaged by the approaching cold winter.

19. Collect seeds from vegetables as you harvest them to use next spring. Store in a cool, dry, insect-free and rodent-free place.

20. Transplant evergreens, ferns, and wild plants.

21. In areas without severe winters, plant fruit trees.

Frost: Mulching protects against light frosts. In northern latitudes, during possible frost season, if it is rainy or cloudy for several days, then suddenly it clears and air becomes still and cold, there is a very good chance of frost. Frosts generally occur in the early morning, just before sun-up. The best way to protect vegetables from frost is to turn on overhead or centrifugal sprinklers just before frost occurs. Or if a frost seems likely, turn sprinklers on before you go to bed and leave them on all night. There are automatic sprinkler systems available commercially that turn water on automatically when the temperature drops to 36°. The water from the sprinklers warms the surrounding earth and air, preventing vegetables from freezing. Having the water running also protects waterpipes from freezing and breaking. Planting on a slight slope lessens frost damage, as cold air is heavier than warm air and will sink downward toward the valley. Severe frost may often cause plant roots to be "heaved up" from the earth. A heavy mulch will often prevent such damage and protect roots should such damage occur.

Cold Nights: Mulch heavily to conserve warmth. Stones in the garden absorb the heat of the day and release it at night, warming the earth. A pond near a garden will keep the surrounding air warmer and add humidity. Watering garden in the early morning warms the earth, inducing plant growth.

Wind: Plant or build windbreaks on the windward side of the garden to protect garden plants. Plant sturdier vegetables where they can stand between the oncoming wind and the more delicate plants. Plant garden on the eastern or southern side of a house or barn (if wind and storms come from the west or north), as the building will provide windbreak.

Wet Season: Provide drainage ditches to prevent garden from drowning. Protect compost from too much water which will wash away valuable nutrients. Protect seeds or young seedlings from being washed away by building a small, removable A-frame or box that may be placed over them during a heavy rain.

Dry Season: Irrigate thoroughly during the dry season and be sure to water deeply. If temperatures are high during the dry season, provide burlap shades for cool-loving vegetables. Water burlap often, as this will provide humidity for the plants. A pond near the garden provides extra humidity.

Heat: Shade all cool-weather vegetables with cloth, burlap, or wooden slats built onto a frame. Provide plenty of moisture for all plants, but do not water when sun is high in the sky. Early morning watering is best.

WEEDS—METHODS OF CONTROLLING THEM

1. Pick them out one by one while soil is moist. Try to remove as much of the main root system as possible.
2. Mulching hinders weed growth.
3. Sprinkle salt on weeds, but not on garden crops.
4. Dig earth between rows and turn it over.
5. Avoid mulches that introduce weed seeds, such as hay.
6. Dig weeds out with a spade.
7. Plant clover or millet. These will crowd out the weeds, as well as add valuable nitrogen to the soil.
8. Remove weed heads before they go to seed.
9. Accept them if they aren't "taking over" and let them be.

CROP ROTATION

As a plant grows, it consumes a portion of the available minerals in the earth. Each type of plant has its own particular mineral needs. To continue to plant the same type of plant on the same plot of ground depletes the natural supply of those specific minerals. Compost and organic fertilizers help somewhat in returning minerals to the soil, but a continuous drain of nitrogen, for instance, may kill soil bacteria and do damage that even compost can't cure.

Crop rotation, which is the practice of following a planting of one vegetable with the planting of a different sort of vegetable provides a resting period for the soil. If one plant removes nitrogen, another plant may be grown in that spot which replaces nitrogen.

Planting the same crop year after year in the same spot encourages the growth of harmful soil fungi, insects, and disease.

The practice of crop rotation often increases yield at harvest time. It promotes healthy growth of plants and their fruit and maintains mineral balance in the soil.

The following is a list of vegetables which may follow previous vegetable plantings in the garden:

    beans—spinach
    Brussels sprouts—carrots, **bee**ts, corn
    cabbages—lettuce, peas, carrots
    cauliflower—legumes
    celery—legumes
    corn—Brussels sprouts
    eggplant—legumes
    legumes—cauliflower, celery, eggplant, potatoes
    potatoes—legumes

GARDEN PEST CONTROL

Conditions that induce disease and mineral deficiencies are the main cause of insect problems. Nature calls out her "destroyers", the insects, to do away with unhealthy plants. Many modern methods work on curing the symptom (insects) rather than the cause (poor plant health), and so a variety of insecticides appear on the market. These poisons do more harm than the insects did in the first place.

If you want healthy, insect-free plants, keep the soil well supplied with proper nutrients and remedy any conditions that may cause disease.

Conditions that may cause unhealthy plant growth are:

- mineral deficiencies
- overwatering
- underwatering
- too much light
- not enough light
- earth too cold
- lack of humidity
- poor soil drainage
- cramped root space
- soil too acid
- soil too alkaline
- sudden temperature changes
- winds and drafts
- too much fertilizer
- chemicals in soil or water
- irrigation water too cold
- uneven watering
- lack of crop rotation
- planting during the wrong season
- lack of mulch
- failure to compost plant remains and destroy diseased or insect-infested ones
- weeds harbor and feed insects, as well as crowd out vegetable roots
- planting unsuitable crops in an area

Know the plants in your garden and their environmental requirements. Group those with similar requirements together, and plan the garden spacing and location so that all the plants will have the environment they need.

There are many beneficial insects and animals that can be brought or attracted to the garden. These creatures will feast on the insects that are destroying your plants. Here is a list of the beneficial insects and animals and how to entice them to your garden.

CREATURE	BENEFIT	HOW TO ATTRACT
Bees	Pollination	Plant flowers and flowering trees and shrubs near the garden.
Earthworm	Aerates and manufactures rich topsoil.	Usually present in most soil or may be imported. Will not live in soils containing chemicals.
Centipede	Eats beetles and roaches	Occur naturally.
Caterpillar Beetle	Eats caterpillars that destroy trees and vegetables	Occur naturally or may be imported.
Lacewing	Eats aphids	Occur naturally or may be imported.
Dragonfly	Eats gnats and mosquitoes	They love ponds and other bodies of still water.
Tachinid Fly	Eats small insects	Occur naturally or may be imported.
Ichneumon Wasp	Eats small insects	Occur naturally or may be imported.
Syrphid Fly	Eats aphids and scales	Occur naturally in most areas.
Ladybug	Eats aphids and scales	May be imported, but often appears naturally where aphids and scales show up.
Praying Mantis	Eats many large and small insects, including grasshoppers, flies, and aphids	Occur naturally or may be imported.

CREATURE	BENEFIT	HOW TO ATTRACT
Spider	Eats small insects	Occur naturally everywhere. Do not kill them.
Gambusia Fish	Eats mosquito larvae	May be imported.
Owl	Eats rodents and insects	Occur naturally. Build owl houses.
Birds	Eat insects, including caterpillars, flying insects, mosquitoes, grasshoppers and borers	Build birdhouses and feeders, birdbaths, pond, grow plants that attract birds, provide building materials such as lint and string, don't let cats run loose or tie a bell on their collars.
Bats	Eat flying insects	They nest in open barns and attics.
Shrew	Eats insects and worms	Occur naturally.
Mole	Eats insects and worms	Occur naturally. Don't kill them—they do not eat vegetables. Mulch to prevent damage due to erosion of topsoil. A mole thumper or anything that vibrates the soil will keep moles away.
Toad	Eats slugs and insects	Pond near garden.
Salamander	Eats insects	Pond near garden.
Lizard	Eats insects	Place a pile of sand near garden and a few large rocks on top of the pile. They Will live there and reproduce as long as pile remains undisturbed.
Small Snakes	Eat insects	Occur naturally.

CREATURE	BENEFIT	HOW TO ATTRACT
King Snake	Eats insects, snakes, and rodents	Occur naturally.
Bull Snake	Eats rodents	Occur naturally.
Bobcat	Eats rodents and rabbits	Occur naturally.
Coyote	Eats rodents and rabbits	Occur naturally.

The following is a list of garden pests and how they may be controlled:

Ants: These creatures are often responsible for the sudden appearance of aphids on trees and perennial shrubs. Ants herd aphids in the same way that man herds cows. Aphids secrete a sort of milk much relished by the ants. Methods of discouraging ants are: Sprinkling salt or bonemeal in their path. Planting mint. Encouraging birds and lizards.

Aphids: Aphids are probably the greatest troublemaker among cabbage, broccoli, and cauliflower, all of which require a very rich soil containing plenty of calcium. Lack of this may have brought them in the first place. The best control is to introduce ladybugs to the infested place. Ladybugs lay small oval-shaped yellow eggs on the undersides of leaves, so watch for them and do not destroy them. Ladybug larvae, which are often mistaken for harmful worms, are also aphid-eaters and should be protected. Another method of controlling aphids is to plant marigolds and nasturtiums in the garden, as aphids dislike their scent.

Caterpillars and Worms: (This does not include earthworms, which are beneficial.) At night, or in the early morning, sprinkle flour on vegetable leaves. Insects will become trapped in the flour which has been moistened by the early morning dew. Another method of control is to introduce caterpillar beetles to your gardens. Birds also help in controlling worms.

Cucumber Beetle: These beetles, which look somewhat like green ladybugs, are not as selective as their name implies. They also love Swiss chard, corn, squash, melons, and lettuce. The best control is to attract birds and lizards, and be certain that soil contains plenty of nitrogen, calcium, and phosphorus. Sprinkle phosphate rock on plant leaves and surrounding soil.

Corn Earworms: They attack corn at the tip where the silk emerges. Be sure soil contains plenty of nitrogen, calcium, and potash. Place a few drops of castor oil on ear tip when ear begins forming.

Cutworms: These creatures "log" small seedlings of plants such as corn, wheat, cabbage, peppers, beans, and tomatoes, by cutting them off at the stem base during the night. To deter them, place a 3-inch-wide strip of stiff paper, encircling plant stem, 1 inch deep into the soil. Cutworms are also deterred by spreading wood ashes around plants. To destroy them, till soil and place chickens in the garden.

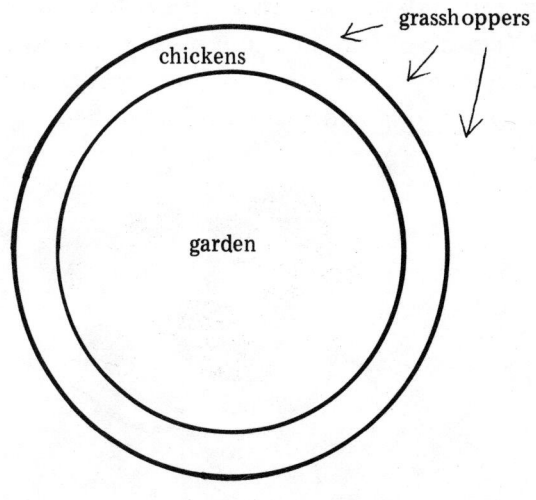

Grasshoppers: Chickens and wild birds are the best method of control for grasshoppers, which ravage gardens through late summer. It is good to have a garden arrangement that allows for chickens to roam space surrounding garden, acting as a sort of "chicken patrol". If you still find many grasshoppers inside the garden, let the chickens in. They prefer bugs to lettuce, but keep a close watch and remove them as soon as they have removed the grasshoppers, or you may find the chickens feasting on your cabbages. Some breeds of chickens are more carnivorous than others. Praying mantis controls grasshoppers also. Tilling soil in fall and spring is another method of control, as this destroys their eggs. Another method of deterring grasshoppers is to spray vegetable leaves with cayenne-pepper - and-water-solution.

Mosquitoes: Do not leave pans of water lying around for mosquitoes to breed in. Attract birds and bats, as they feed on mosquitoes, as do large insects. In ponds, fish—especially Gambusias—will take care of mosquito larvae.

Snails and Slugs: These insects destroy many plants, such as flowering shrubs and strawberries. Methods of control are: Trapping them in a dish of beer. Sprinkling a border of sand, wood ashes, or lime around the garden. Attracting birds and toads to the garden.

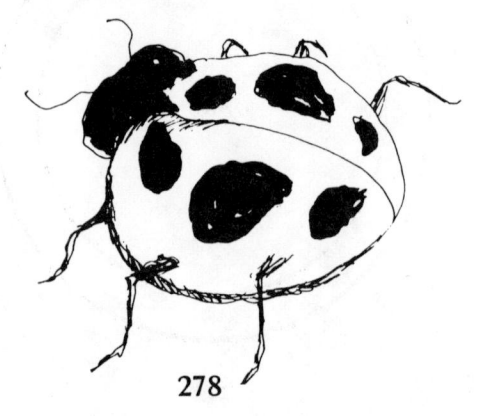

ORGANIC INSECTICIDES AND DETERRENTS

1. Mix a few tablespoonfuls of cayenne pepper and a quarter cup of biodegradable soap, in a pail of water. Use this solution to spray leaves of the plant.

2. Grind mint, aromatic marigolds, nasturtiums, and daisies (leaves and flowers) and soak in a pail of water for a day or so. Spray plant leaves with the liquid extracted from this solution.

3. Plant any of the following, as most harmful insects dislike the scent they produce: mint and other aromatic herbs: absinthe, anise, asters, basil, chrysanthemums, cosmos, geraniums, larkspur, marigolds, mugwort, nasturtiums, sage, silver king sage, silver mound, southernwood, tarragon, wormwood.

4. Sprinkle bonemeal or bloodmeal around plants.

5. Mix a cup of wood ashes and a cup of lime in 2 gallons of water and spray leaves with this solution.

6. Mix 1 part hardwood sawdust and 1 part wholewheat bran to 3 parts molasses-and-water and spread this solution around the base of each plant to trap worms and other insects.

7. Pine sawdust deters insects.

8. To trap yellow jackets and other bothersome flying insects, fill a narrow-necked jar half full with water and add a little honey, molasses, or fruit juice. Insects will find their way into the bottle but the narrow neck will prevent them from finding their way out.

MAMMALS

The diets of various mammals include:

Bats: Flying insects.

Beavers: Tree bark, young saplings.

Chipmunks: Seeds, berries, grain, nuts, acorns, bird eggs, baby birds, mice, insects.

Gophers: Wild greens, fruits, and vegetable roots.

Ground Squirrels: Garden vegetables, as well as seeds, nuts, acorns, berries, grain.

Mice: Young seedlings, nuts, grain, seeds.

Moles: Worms, insects. Contrary to popular belief, moles do not harm vegetable roots or plants, except that they may displace young seedlings through their burrowing, or cause earth surface to dry out. Moles aerate the soil.

Porcupines: Tree bark, vegetables, tool handles.

Rabbits: Young seedlings, tender greens.

Tree Squirrels: Seed, insects, nuts, acorns, grain, and berries.

DETERRING RODENTS

1. Mulching deters or prevents damage from some burrowing rodents.
2. Attract snakes, hawks and other natural predators.
3. Place an inverted bottle into the hole of a burrowing rodent. The reverberations and echoes will frighten him away.

4. Plant Scilla flower bulbs around the garden.
5. Build a thumper.
6. Construct a rodent-proof fence (as mentioned in paragraphs on The Fence).

THE ORCHARD

COMBINATION GARDEN-ORCHARD PLAN

The plan shown in the accompanying diagram provides for spring and fall vegetable gardens, a fruit tree orchard, and a berry vineyard. It also provides space for a chicken house.

The chickens may be allowed to range-feed outside the garden area, acting as a "chicken patrol" against invading insects, or they may be kept inside any of the four gardens to rid an area of insects and "till" last season's garden spot. They till the earth by scratching the earth, ridding it of harmful insects, adding manure, and scavenging through compost remains.

Alfalfa may be planted in the fruit tree orchard and chickens allowed to range-feed on the alfalfa. This will add alfalfa remains and manure to the orchard site and rid the area of insects.

This plan also allows for a storage shed to store garden tools, organic fertilizers, and chicken feed. The shed also provides a windbreak against north winds.

When planting the garden and orchard, plan to place the tallest vegetables and trees on the north side of the plot.

N

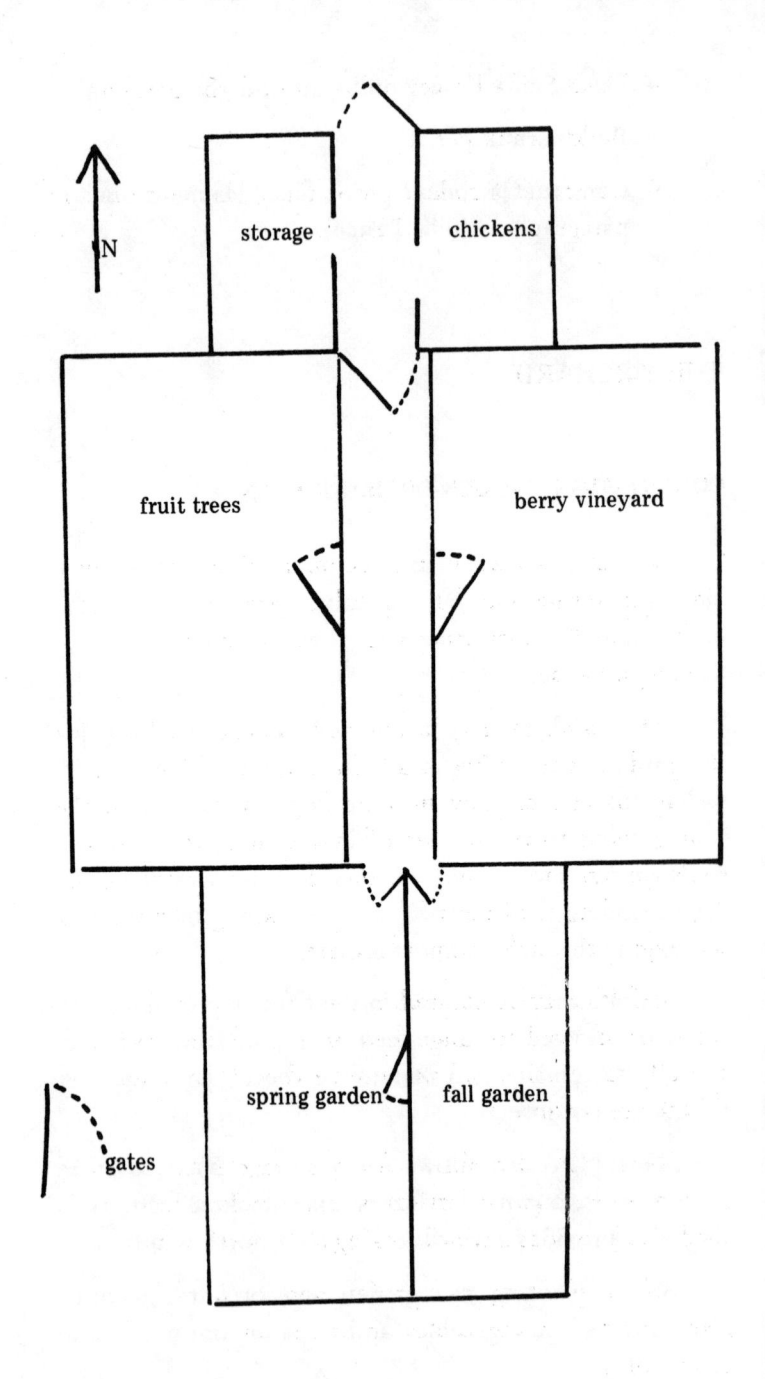

storage | chickens

fruit trees | berry vineyard

spring garden | fall garden

gates

FRUIT TREES STARTED FROM SEED

Some fruit trees do very well when started from seed, and in some areas will thrive and grow better than trees that have been grafted or transplanted. By starting a tree from seed it has more of a chance to become acclimated to conditions of the area. It is best to experiment with several varieties, and if possible, to use seeds from fruit-bearing trees in the vicinity.

The only drawback of starting trees from seed is that it takes additional time to grow a tree from seed to maturity.

Fruit trees that may be started from seed include:

	Years to Fruit Bearing Age
almond	7-8
apple	5-10
apricot	3-4
avocado	5-7
cherry	4-7
peach	4-6
pear	6-8
pecan	6
plum	4-5
walnut	8

Most fruit trees, except tropical and citrus, require a period of freezing before the seed will germinate. For this reason, it is best to start seeds directly into the ground outdoors during the winter in an area that can be watched and carefully tended. If you are in an area without freezing periods during the winter, this period may be simulated by placing seed in flats filled with sand in a freezer for 3-4 months before planting. Almonds and apricots need be frozen for only one month before planting.

Seed is planted into the ground ½ to 1 inch deep, depending upon the size of the seed. Space plants 1 to 2 feet apart if they are to be transplanted when 1 or 2 years old. If you do not wish to transplant them later, space seeds 20 to 30 feet apart in all directions. Soil should be a deep, rich, well drained loam and the seeds and young trees should be watered regularly.

It is best to keep a close eye on the fruit trees for the first few years to protect them from frost, deer, livestock, and a variety of other things. Once you get the plants established, they have a much better chance of survival.

Avocado seeds may be germinated by slicing a thin layer off of the top and bottom of the seed and placing it in a pot with drainage holes filled with moist sand. Keep continuously moist until it germinates, 1 or 2 months later.

Citrus trees are delicate and should be started from commercial plants. A frost-free environment is necessary for them to grow.

FRUIT TREES FROM CUTTINGS

Some fruit trees may be propagated by cutting off branches or suckers in the late fall or early spring and placing them in a bucket of wet sand.

Spray the branch with a mist of water several times a day and keep it in a warm, humid, sheltered spot.

The following may be started in this way:

apples	olives
berries	plums
currants	pomegranates
figs	quinces
grapes	

PLANTING YOUNG FRUIT TREES

Deciduous-leafed fruit trees should be planted during their dormant stage, which begins when they have dropped their leaves in the fall, and continues until early spring when new leaf buds begin to develop.

1. Test the soil. Check pH and the nutritional requirements for the tree you are planting. If the soil is too acid, add bonemeal, wood ashes, or limestone. If the soil is too alkaline, add leaves, compost, or wood chips.
2. Dig a space 3 feet wide, 3 feet deep.
3. Make a mixture of sand, compost, silt, topsoil, and leafmold. Do not use manure or anything that encourages rapid growth to this mixture, because the top will develop more quickly than the roots and the plant will become top-heavy.
4. Fill space 1/3 full with above mixture.
5. Throw in a few large stones.
6. Fill space another 1/3 full with topsoil.
7. Prune tree roots, cutting off dead and broken roots to encourage growth of feeder roots.
8. Throw a few hundred earthworms in the space.
9. Place tree in the center of the space and a stake in the earth next to the tree trunk to help tree grow straight for the first year or so.
10. Fill in around tree with above mixture, pressing the earth in gently but firmly around the tree.
11. Water thoroughly to settle the earth.
12. Mulch with 2 inches of compost, then 3 inches of leaves or other mulch, then 1 inch of gravel or small stones.

Heavy Clay Sloping Land

large rocks — sandy mixture — clay — 3 feet — gentle slope

Heavy Clay Porous Subsoil

clay — mixture 3 feet or more — porous subsoil

POROUS SOIL

Wet Climate

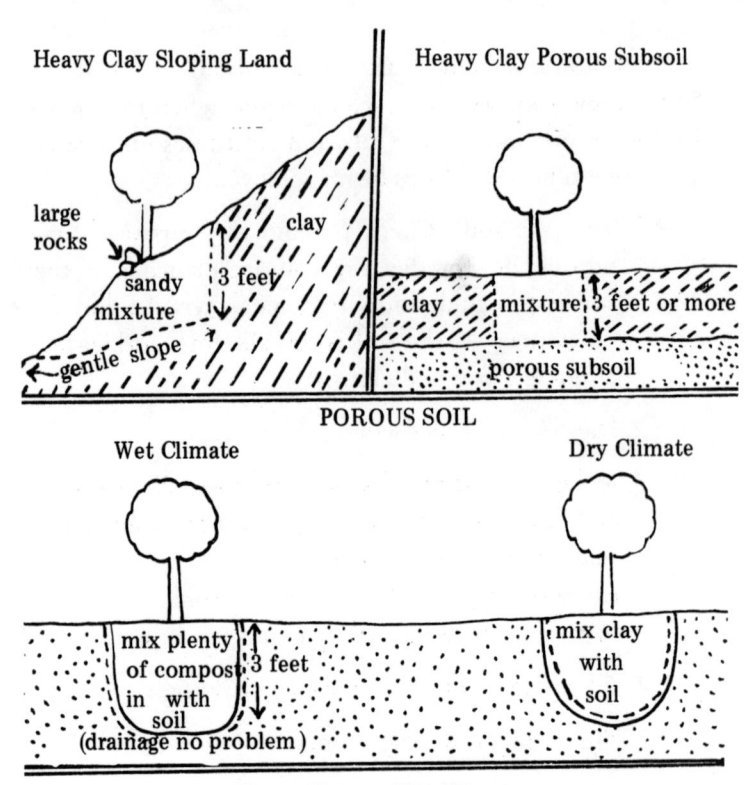

mix plenty of compost in with soil — 3 feet — (drainage no problem)

Dry Climate

mix clay with soil

Heavy Clay — Wet Climate

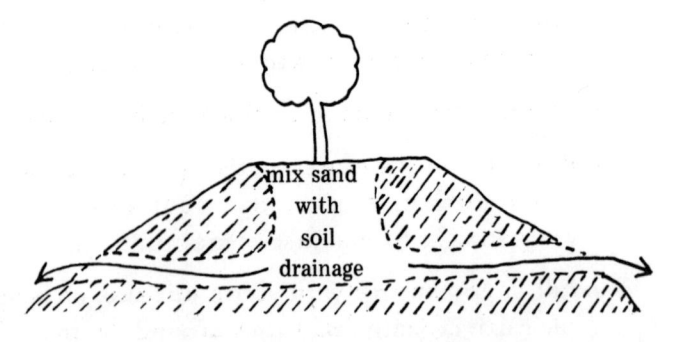

mix sand with soil — drainage

When planning the orchard, be sure to space the trees far enough apart that they won't interfere with sunlight or root growth. Fifteen feet is the minimum distance between trees. Twenty to thirty feet is better.

Place the tallest trees on the north side of shorter varieties:

Tall	Medium	Short
cherry	apple	apricot
date palm	avocado	citrus
	pear	fig
		peach
		plum

Birds may be a pest in the orchard, as they are often attracted to the fruit. Planting and cultivating a few wild berry vines and berry shrubs, will draw them to those instead, as they prefer the taste of wild, bitter fruit to the sweet taste of cultivated fruits and berries.

FRUIT TREES

APPLE

Soil pH: 5-6.

Nutritional Requirements: nitrogen and potassium.

Cultivation: Plant new trees in the spring or fall. Fertilize each year just before tree blooms. Water sparingly, about once every 3 weeks during the dry season. Hold off water in fall.

APRICOT

Nutritional Requirements: moderate: potassium.

Cultivation: Plant new trees in very early spring. Fertilize only when tree shows symptoms of deficiency. Needs plenty of water during the dry season. In areas where late spring frost is a problem trees should be planted on a north slope. The soil on a north slope remains cold late in the spring keeping the trees dormant much longer. When trees are dormant they are not damaged by frost. In some areas hard freezing in the late spring is common. When the sap is frozen the tree is severely damaged or killed. This condition can be prevented by north slope planting.

AVOCADO

Soil pH: 6-8.

Nutritional Requirements: phosphorus and nitrogen.

Cultivation: Plant new trees in the early spring. Protect from severe winters until they are 3 years old. Fertilize during spring and summer. Water frequently all year round. It will be necessary to have a grafted tree (from a nursery) or at least 2 trees near each other for pollination and fruit formation to take place.

CHERRY

Soil pH: 6-8.

Nutritional Requirements: nitrogen.

Cultivation: Plant new trees in fall except in areas of cold, severe winters. Mulch heavily. Provide water during the dry season. When harvesting, pick the stem as well as the cherry, as this will help tree to bear more fruit next year.

CITRUS

Soil pH: 5-6.

Nutritional Requirements: nitrogen and phosphorus.

Cultivation: Must be grown in a frost-free area or in a greenhouse. Should have full sun and protection from dry winds. Fertilize once a year. Mulch heavily. Keep continuously moist just after planting and just after the tree is established; water every few weeks. As tree matures, water less, and avoid "drowning" it with too much water.

FIG

Soil pH: 6-8.

Nutritional Requirements: minimal.

Cultivation: Start new trees in the early spring. Soil must have good drainage and plenty of humus. Mulch heavily. Provide fertilizer only when plant shows symptoms of deficiency, as over-fertilizing will cause much leaf and little fruit development. Provide abundant water.

PEACH

Soil pH: 6-8.

Nutritional Requirements: moderate.

Cultivation: Start new trees in the early spring. Requires cold winters and will withstand hot summers. Fertilize only in the early spring. Soil must be a loose, well drained loam. Water moderately.

PEAR

Soil pH: 6-7.

Nutritional Requirements: lime.

Cultivation: Plant new trees in the very early spring, or in the fall in very mild areas. Prefers clay-type loam, well drained. Protect from intense sun by planting in a cool spot. Supply plenty of water throughout the year.

PLUM

Soil pH: 6-8.

Nutritional Requirements: potassium and phosphorus.

Cultivation: Plant new trees in late winter. Soil should be rich in humus and well drained. Supply abundant moisture.

FERTILIZING FRUIT TREES

Take care not to over-fertilize your fruit trees. The simplest and most natural way is to surround the tree with a mixture of compost humus, well rotted manure, and any mineral (such as phosphate rock) that the tree requires. Rain or irrigation water will carry the nutrients down to the feeder roots. Digging fertilizer into the soil often damages these delicate roots. Fertilizer should be placed at the drip line, which is an imaginary line perpendicular to the earth, extending as far as the outermost branches. Nutritional deficiency symptoms, mentioned earlier in this cahpter, apply also to fruit trees.

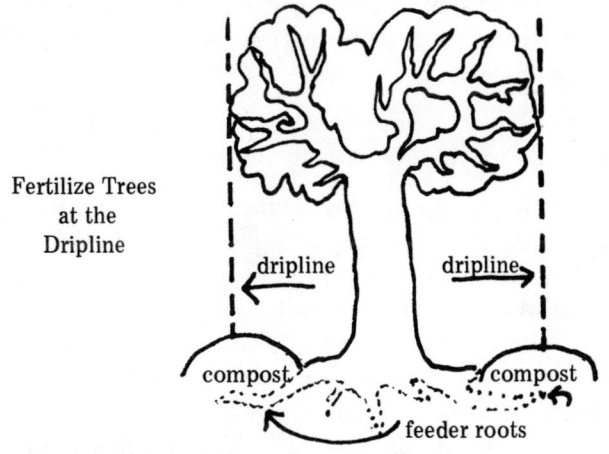

Fertilize Trees at the Dripline

PLANTING GRAPE AND BERRY VINES

Grape and Berry vines are usually planted in rows with some means of support, such as a trellis or fence.

Dig a trench 2 feet wide and fill half way with a mixture of 1 part sand or silt, 2 parts compost humus, and 2 parts garden loam. Place grape or berry vines 3 feet to 6 feet apart (depending on how much each vine is expected to spread) in the trench and fill trench with remainder of the mixture. Water thoroughly and mulch lightly around the base of each plant. As plants begin to grow, train them to climb the trellis.

Plants may be procured from a nursery, or propagated by cuttings. To propagate by cuttings: In fall or spring, cut runners or suckers from a healthy plant. Place in a bucket of wet sand in a warm, sheltered area and spray often to keep atmosphere humid. When roots develop, they are ready to plant.

Some plants grow runners that already have roots which develop in small clumps along the length of the runner. These may be planted directly in the garden. Procuring nursery plants is often preferable over starting runners, as 1- or 2-year-old plants will produce fruit to be harvested much sooner.

The first season after planting the vines, remove all flowers to prevent fruit formation, as this will strengthen the root systems.

Prune plants when necessary to prevent the runners from becoming an uncontrollable tangle.

Fertilize grape and berry vines annually, in the early spring. Make a mixture of 3 parts compost humus, 1 part well rotted manure, and 1 part mineral rock or organic substance (such as phosphate rock, bonemeal, wood ashes). Be careful not to add any substance that will alter soil pH to one which is incorrect for the vine you are planting. Many types of vine, such as blueberry, are very particular about their soil pH, and the proper one should be maintained. Most berries require an acid-type soil and the addition of too much limestone, for instance, would make the soil too alkaline.

The following is a list of plants and the time it takes from planting a runner or small plant to harvest of the first crop.

blackberries	2-3 years	gooseberries	2-3 years
blueberries	5-8 years	grapes	2-4 years
cranberries	3-4 years	raspberries	2-3 years
currants	3 years	rosehips	1-2 years
elderberries	1-3 years		

BERRIES AND GRAPES

BLACKBERRY, BOYSENBERRY, LOGANBERRY

Soil pH: 5-7.

Nutritional Requirements: nitrogen, potassium, and phosphorus.

Cultivation: Start new plants in the late fall or very early spring. Mulch heavily. Irrigate thoroughly in the summer, as they need much water, especially while the fruit is ripening. May be planted on a north slope, as they prefer a cool climate while ripening.

BLUEBERRIES

Soil pH: 4-5.

Nutritional Requirements: Add only compost and high-acid humus. Never add lime or anything that may have an alkalizing effect.

Cultivation: Start new plants in the fall or early spring. Mulch heavily with a high-acid mulch. Needs abundant moisture at all times.

CRANBERRIES

Soil pH: 3-4.

Nutritional Requirements: Minimal. Add compost each spring.

Cultivation: Start new plants in the early spring. Requires fairly mild year-round climate. Prepare an area that can be flooded as needed for irrigation. In the summer, flood with water whenever the soil dries out. Just before the freezing period, flood with water to cover the plants. Drain in the spring.

CURRANTS

Soil pH: 6-8.

Nutritional Requirements: nitrogen and potassium.

Cultivation: Start new plants in the late fall or winter in a well drained clay-type soil, in a shaded spot or on the north slope. Requires much moisture during the summer. Hardy in a cold climate.

ELDERBERRY

Soil pH: 6-8.

Nutritional Requirements: nitrogen.

Cultivation: Start new plants in the fall or very early spring. Requires a cool, moist habitat, preferably near water, and a high humidity.

GOOSEBERRY

Soil pH: 6-8.

Nutritional Requirements: potassium.

Cultivation: Start new plants in the fall. Mulch heavily and provide plenty of water. Requires a cool environment, especially when fruit is developing, and may do well on a north slope.

GRAPE

Soil pH: 6-8.

Nutritional Requirements: nitrogen, phosphorus, and potassium.

Cultivation: Start new plants in the early spring in cold areas, in the late fall in mild areas. Choose a variety of grapes suited to your climate. Provide a well drained, warm location in full sun. Moderate water requirements.

RASPBERRY

Soil pH: 5-6.

Nutritional Requirements: plenty of compost humus.

Cultivation: Start new plants in fall or early spring. Soil should be well drained and rich in humus. Abundant moisture should be supplied during dry season.

ROSEHIPS

Soil pH: 5-6.

Nutritional Requirements: moderate.

Cultivation: Start new plants in fall or early spring. Provide a rich, well drained humus and mulch lightly. Fertilize each spring with a side-dressing of compost humus. Plants are hardy.

GRAPE AND BERRY ARBOR

Dig two parallel trenches 2 feet wide and 8 feet apart. Fill with river silt or well drained humus.

Build a trellis next to each trench for the vines to grow on.

Place grape or berry plants 6 to 8 feet apart in the trench.

Between trenches, place a mounded row of raw compost wastes on top of the soil.

Run a water line which has been punctured every few inches to allow small drips of water to fall on the compost.

Let this drip run continuously throughout growing season.

Replenish compost when needed.

This method takes care of fertilizing and watering the grape or berry plants all at the same time.

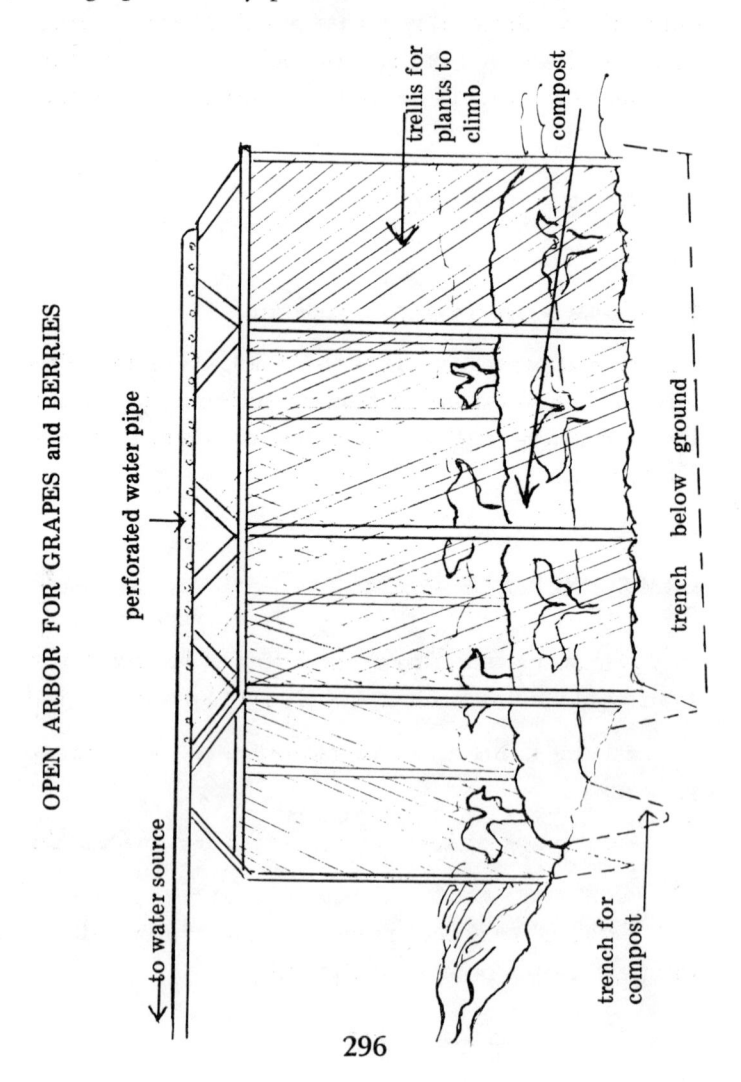

OPEN ARBOR FOR GRAPES and BERRIES
- trellis for plants to climb
- compost
- perforated water pipe
- to water source
- trench below ground
- trench for compost

THE HERB GARDEN

Herbs are beautiful, medicinal, tasty, fragrant, and flower-bearing. They are easy to cultivate and are beautiful as a border around the garden or in a garden area of their own. Some do well as house plants.

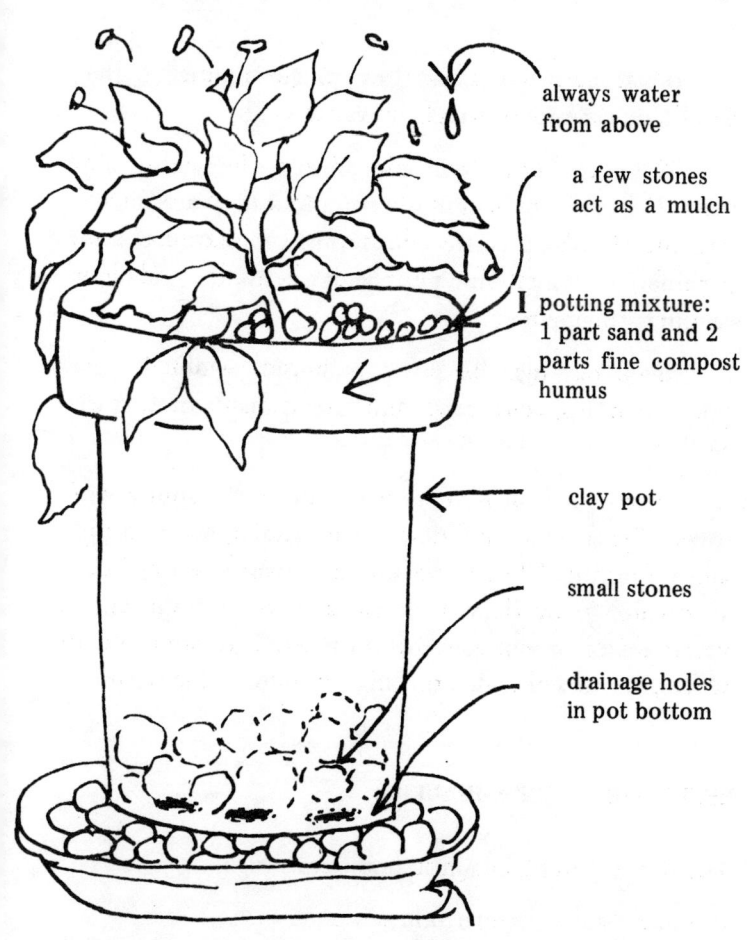

always water from above

a few stones act as a mulch

potting mixture: 1 part sand and 2 parts fine compost humus

clay pot

small stones

drainage holes in pot bottom

Place on a tray containing sand or small stones. When plant is watered, moisture seeps down through drainage holes onto the tray. Stones help retain the moisture and allow it to evaporate slowly, providing constant humidity. Water plants when stones in bottom tray become dry. Rinse entire plant under a faucet once a week.

Cultivation

Prepare soil in the spring by tilling 1 to 2 feet deep and adding silt or sand to the soil if you are in a heavy-soil (such as clay) area. Add a good quantity of compost humus and mound earth up as done for vegetable cultivation. This provides good drainage and aeration of the soil.

Most herb seeds are slow to germinate, so they should be soaked overnight in warm water.

The next day, drain and air-dry the seeds. Sow them ¼ to ½ inch deep in the soil mound. Water lightly and mulch. Keep earth continuously moist until seeds germinate, at which time push back the mulch and allow seedlings to emerge.

Seed may also be sown in indoor seedbeds, peat pots, or in a green house, and later transplanted to the garden.

Herbs may also be grown indoors in a sunny window. Be certain that they receive adequate moisture and humidity. They prefer a cool to warm temperature, so do not grow them too near a stove or fireplace. A water mister or sprayer is often needed to spray plants with water to provide the high humidity they require.

HERBS — NATURE'S HEALERS

Bee Stings: Mud and garlic poultice.

Stomach Ache: peppermint, ginseng.

Vitamins: red raspberry—A,B,C,G,E; parsley—A,C; alfalfa—A,B,E,K,U.

Memory: rosemary, Gotu Kola.

Mind Rejuvenation: Gotu Kola.

Smoking Substitute: equal parts of red raspberry, red clover, coltsfoot, chamomile.

Stimulants: peppermint, ginseng, Gotu Kola, cayenne, guarana.

Sedatives: hops, chamomile, catnip, skullcap, motherwort, vervain, valerian.

Menstrual Cramps: black cohosh, red raspberry, ginseng.

Body Rejuvenation: ginseng.

Hair Rinse: jabora, henna, rosemary, sage, chamomile.

Blood Purifier: Gotu Kola, ginseng, golden seal.

Cure All: Golden seal.

Constipation: licorice root.

Sore Throat: slippery elm, sage, licorice root.

Colds: garlic, peppermint.

Poison Oak or Ivy: EXTERNAL washes to relieve itch: sassafras + goldenseal or Lobelia + jewelweed or Indian soaproot.

Diseases of Teeth and Gums: chew raw garlic before retiring at nighttime.

COLDFRAMES AND GREENHOUSES

Coldframes and greenhouses are structures used to shelter plants and seedlings from extreme weather and provide a humid, protected place for starting or growing plants. Coldframes and greenhouses work on the principle of allowing the sun's light rays to enter, changing them to heat rays, and preventing their escape.

A coldframe is a bottomless, narrow frame, equipped with a hinged, transparent lid, such as glass or fiberglass. It is often desirable to have a shade and slatted cover which may be interchanged with the glass cover to provide shade or partial shade for some plants (such as lettuce) during the hot months.

A hotbed is identical to a coldframe, except that it also has a means of heating from below. This allows seedlings and plants to be set out earlier (in the hotbed) in the season and provides warmth for growing plants late in the season and in areas with cold nights. Heat may be supplied by an underlayer of decomposing organic material, such as fresh manure or raw compost, or by using electric heating cables available for this purpose. When using the organic material, place a 2-foot-deep layer of the material inside the frame and a 6-inch layer of soil on top of it. Mound earth up around the outside of the frame to help retain heat. Seeds are then planted in the top layer of soil. Be sure to provide plenty of water and fresh air, as the heat rising from below may dry the earth and air within the frame. Open lid often for ventilation.

Greenhouses are generally larger than coldframes, often the size of small houses, with glass or other transparent material used for the roof and siding. A greenhouse is invaluable for propagating plants, starting seedlings, and growing vegetables during the cold months, as it provides a warm, humid atmosphere which induces healthy plant growth. Vegetables grown in a greenhouse will be more tender and sweet tasting than those exposed to the outside weather. Growing under protection, they don't need to toughen their skins in order to withstand the elements. These vegetables also tend to be less hardy, and precautions must be takent o prevent insects and disease which often plague the greenhouse:

1. Allow for proper ventilation.
2. Remove unhealthy plants immediately.
3. Destroy insect-infested plants.
4. Do not over-water.

5. Protect plants from drafts.
6. Regulate amount of fertilizer (organic!) carefully.
7. Never smoke in a greenhouse.
8. Never use chemicals of any kind in a greenhouse (or anywhere, for that matter).

The greenhouse may be built any size or shape. It should have a southern exposure free of obstruction, such as trees, which may block out light during the winter months.

The floor may be of earth or concrete. The frame should be built of durable wood, such as redwood, or nonrusting metal. Walls should be constructed of glass or another transparent material, such as fiberglass. Provide windows which open for ventilation, but are adjustable so that they will not cause drafts. Planting beds about 2 feet high are built into the greenhouse with adequate space between beds to walk between them freely. It is good to install a subsurface water system in each bed to provide irrigation. Be careful not to over-water, and always water in the morning.

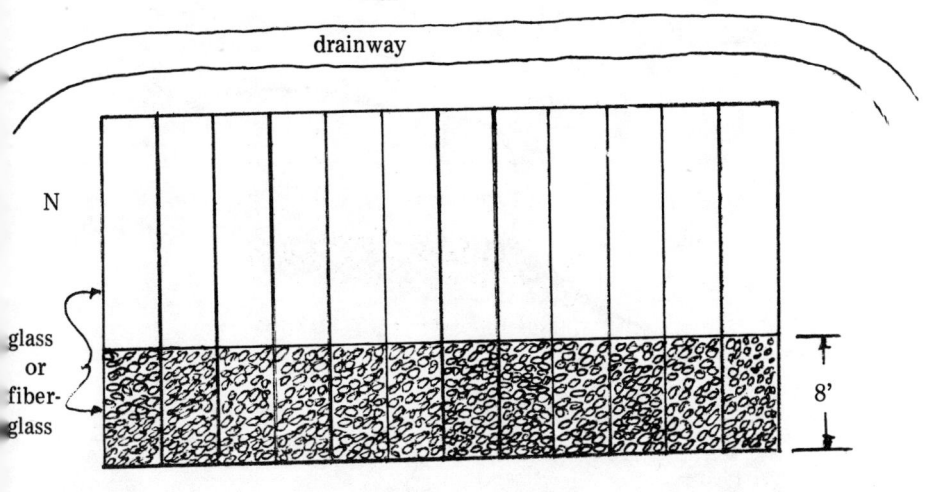

A means of heating the greenhouse should be provided. This may be either woodstove, electric, gas, or solar.

BUILDING A SOLAR GREENHOUSE

The best place to build a Solar Greenhouse is into a hillside that slopes downward to the south. If you do not have a south-sloping hill, one can be created by pushing up a ridge of soil with a bulldozer or other dirt-moving equipment. The ridge of soil must be thoroughly packed and leveled to the proper contour.

Stake out the perimeter of the greenhouse and use a bulldozer, loader, or other dirt-moving equipment (it may also be done by hand) to trim the hillside and level the floor area. The floor area should slope about ½ inch to the foot downward toward the south. Provide a drainway so that water from the hill drains around and away from the building. If the hillside is not stable, this type of construction is not recommended.

SOLAR GREENHOUSE IN HILLSIDE

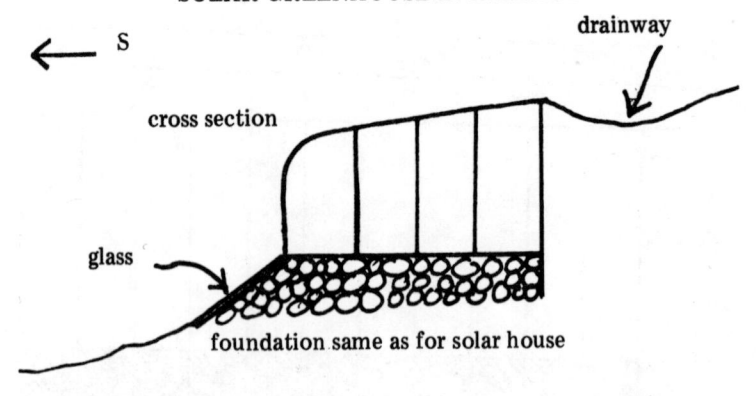

← S

drainway

cross section

glass

foundation same as for solar house

See also in Chapter II, Using Nature's Energy—Solar Home.

IN CONCLUSION

In order to survive, man must return to the land. Only then will he find the mental, physical and spiritual outlet his body was designed for. The transition is not an easy one, and for many it will never be realized. But for those who can overcome the obstacles, the rewards will be great. A half century of unnatural living has left man confused, distorted mentally, weakened physically, broken spiritually, and with a polluted world. Man is also plagued with chemicalized food, inflation, and an energy crisis. Those who must remain in urban areas will find food and other products becoming even more chemicalized, an upward spiraling of prices, and many inadequancies. Urban areas do play a very important part in the development of culture and civilization, but a change in mind, body and spirit must occur if the city is to live.

There exists a vast storehouse of free energy and an unlimited food potential for those with the energy and the initiative to act creatively. There is much land left to work with, in this country as well as in other parts of the world. There is an ascending trend toward group occupancy of land. This is a great idea and can be very successful if properly planned. Most failures in communal homesteads are caused by incompatibility and lack of a common denominator, such as a leader, spiritual practice, or garden development. Enthusiasm and other positive attitudes certainly help to make one's community or one's life more creative and fulfilling.

We hope that this book will be a great benefit to all beneficial life.

Howard & Suzen

INDEX

acid soil, 184
acid-loving plants, 51,52,184
adders tongue, 17,18,35
adobe, 88
alder, 35,133,134
alfalfa, 52,134,185
algae, 156
alkaline loving plants, 52,185
Allegheny spurge, 134
allergies, 64
almond, 283
American lotus, 158
animals, beneficial, 274
ants, 276
aphids, 274,276
apple, 283,284,287,288
apricot, 283,287,288
arbors, 295,296
arborvitae, 134
arrow arum, 35,158
arrow head, 158
artichoke, 197,200,218
ash, 17,19,134
ashes, 181,182,255,292
asparagus, 197,201
avocado, 287,288
azalea, 17,24
azolla, 158

basalt, 46
basement, 88
bats, 275,280
batteries, 118,120,124,127
beach grass, 133
beams, 98,99,102
beans, 134,197,198,202,239
beavers, 280

bedrock, 15,139,153,157,161
bees, 274
bee stings, 298
beetles, cucumber, 277
beets, 197,203
beggarweed, 134
beneficial animals, 274
bentonite, 157
berries, 284,291-296
birds, 275
blocking, 99
bloodmeal, 181
blue clay, 53
blue flag, 35,158
bobcat, 276
bonemeal, 181
boron, 40,255
box elder, 134
brake fern, 134
broccoli, 197,204
brome grass, 133
Brussels sprouts, 197,205
buckwheat, 134

cabbage, 197,206
cabbage palmetto, 17,25
cabomba, 158
calcium, 255
cantaloupes, 197,207
Cape pondweed, 158
carrots, 197,208
caterpillar beetle, 274
caterpillars, 276
catfish, 158
cattails, 17,20,158
cauliflower, 197,209
cedar, red, 133

304

INDEX

celeriac, 197,210
celery, 197,210
cement, 108,144
centipede, 274
centipede grass, 134
cesspool, 106
chainmill, 84
chainsaw, 84
chard, 197,211
cherries, 283,287,289
chickens, 177,277,281
chickweed, 35
chipmunks, 280
citrus, 287,289
clams, 158
clay, 47,139,155,156
clearing land, 84
climate, 56
clover, 134
coldframes, 299
colds, 299
collards, 197,212
Colorado blue spruce, 134
coltsfoot, 133
companion planting, 188
compost, 177-181,261
   Superlative Compost Developed by Orem's Organic Gardens, 256
concrete, 88,144,167
converters, 119
coontail, 158,253,256
corn, 197,213
corn earworm, 277
cottonwood, 134
cover crops, 262
cowlily, 158

cranberries, 293
crayfish, 158
creek, 150
creek well, 153
creeping nettle, 134
creeping willow, 133
crested woodfern, 17,21
cucumber beetles, 277
cucumbers, 197,214
culverts, 76
currants, 284,294
cuttings, 196,284
cutworms, 277

dams, 136,165,167
daphnia, 159,256
date palm, 287
deed, 70
deficiency symptoms, 25ⁿ 260
deposit, 69
dikes, 15
dogs, 66,67
dolomite, 186,255
doors, 104
Douglas fir, 134
dragonfly, 274
drainfield, 106,109
duckweed, 158
dust storms, 61

earthquakes, 61
earthworm, 274
earworm, corn, 277
eggplant, 197,215
electricity, 115-127
elephant's ear, 158

INDEX

elm, 134
elodea, 36,158
endive, 197,216
energy, 116
erosion, 78,133-137
escrow, 69

fences, 111,175
ferns, 17,21,26,29,30,35,134
ferrocement, 144
fertile soil indicators, 50
fertilizer, 254,261
fescue, 134
figs, 284,287,289
firebreaks, 137
fir, Douglas, 134
fires, 61,132,137
firewood, 132
fish, 154
fish ponds, 154
flats, 193
flies, 67,274
floors, 97
flowering rush, 36,158
flowers, 259
footing, 89-93
forms, 92,107,112
foundation, 89-93
frame, 98
frost, 270,288
fruit trees, 282,285,288

gambusia, 158,275,278
garden, 173,178,277,282
garlic, 197,217
generator, 124
ginseng, 298,299

golden club, 36,158
gooseberry, 36,134
gophers, 280
Gotu Kola, 298,299
grapes, 292,294,295
grasshoppers, 277,278
gravel, 140
gray birch, 133
greenhouses, 299-302
ground squirrels, 280
groundwater, 7-16,140,141
growth chart, 197
gullies, 136

hackberry, 134
hawthorn, 133
heather, 134
heating, 127-132
hedge hyssop, 17,22
hedges, 111,175
herbs, 297,298
honey locust, 134
honeysuckle, 134
horsetail, 17,23,65
hotbed, 300
houses, 87,128
huckleberry, 37,133
humidity, 58,271,297,298
hurricanes, 60

ichneumon wasp, 274
insects, 273-280
iron, 40,255
irrigation, 176
ivy, 37,65,134

Japanese spurge, 134

INDEX

Jerusalem artichokes, 197,218
joint fir, 134
joists, 97

kale, 197,219
Kenilworth ivy, 37,134
kohlrabi, 197,220

lacewing, 274
ladybug, 274
land, 9,73
land leveling, 85
land slopes, 41,42,43,115
land stability, 82
larkspur, 65
leaves, 252
leeks, 197,221
legal aspects, 67,69,83
legumes, 134
lentils, 134
lettuce, 197,222
lily turf, 134
limestone, 186,253
lime loving plants, 185,290
lizard, 275
lizard tail, 158
loam, 48
locust, 134
logs, 88
lupine, 65,133
luzula, 134
lyme grass, 134

magnesium, 255
manganese, 40,255
manure, 261
maple, vine, 133,134

marsh fern, 17,26,37
marsh marigold, 158
materials, 87
meditation, 71
mice, 280
microclimate, 62
millet, 134
mole, 275,280
moon, 189
mulching, 251
mosquitoes, 158,278
mushrooms, 65,197,223
mustard, 224
myriophyllum, 158

neighbors, 66,83
nettle, 37,65,134
nitrates, 40
nitrogen, 254
Norway spruce, 134

oak, 40,134
okra, 197,225
olives, 284
onions, 197,226
orchard, 281
outcrop, 16,140
outhouse, 105
owl, 275

pachysandra, 134
palm, 17,25,158,287
palmetto, 17,25
parsnips, 184,197,227
parsley, 184,298
peach, 287,290
peanuts, 197,228

INDEX

pear, 290
pearlwort, 134
peas, 134,198,229
peat moss, 252
pecan, 283
peppermint, 298
peppers, 198,230
pesticides, 40
pest plants, 65,271
phosphate, 181,255
phosphorus, 254,257-260, 293,294
pickerel weed, 158
piers, 95
pignut, 133
pine, 133,134
pipe, 141,142
planting, 187-191
plastering, 144
plum, 284,287,290
poison, 12,13
poison ivy, 65
poison oak, 65
poison sumac, 17,28,38,65
poisonous plants, 65
pollination, 213
pollution, 11,80
pomegranates, 284
ponds, 154
poplars, 134
porcupines, 280
potassium, 255-260
potatoes, 198,231
potatoes, sweet, 198,232
praying mantis, 274
precipitation, 63,139
privy, 105

pumps, 168

quinces, 284

rabbits, 280
radishes, 198,233
rafters, 100,101
rain, 62,63,139
rainshadow, 63
rainwater, 159
rape, 134
rattlesnake fern, 17,29
recharge reservoirs, 163
red cedar, 133
regulator, 118
reservoirs, 159-170
rhubarb, 198,234
rice, 198,235
river, 135-139
roads, 55,74
rock formations, 46
rocks, 46,115
rodents, 280
roof, 100
roots, 257,285,291
rosehips, 292
rutabaga, 198,237
rye, 134

salamander, 275
salt, 40,271
sand, 140
sassafras, 133
sawdust, 253
scales, 274
scrub oak, 133
seaweed, 253

INDEX

seeds, 191,283
septic tank, 106
sesame, 198,238
shrew, 275
silt, 49
skunk cabbage, 17,31
slugs, 275,278
snails, 158,278
snakes, 275,276
snowstorms, 60
soil, 45,181-185
soil composition, 47,53
soil tests, 53,54,183
soil types, 45
solar heat, 95,126-129
soybeans, 134,198,239
spiders, 275
spignate, 17,32
spike rush, 38,133,158
spillway, 165-167
spinach, 198,240
spinach, New Zealand, 198,241
springs, 14,146
spruce, Colorado blue, 134
squash, 198,242,243
squirrels, 280
stomach ache, 298
stone, 88,115,253
strawberries, 198,244
streambank erosion, 134,135
sulphur, 255,257,258
sun, 79,126
sunflower, 198,245
sweet fern, 134
syrphid fly, 274

tachinid fly, 274

tapegrass, 158
terracing, 85
thunderstorms, 60
toad, 275
tomatoes, 198,246
topsoil, 17,140
tornadoes, 61
trace minerals, 255
transplanting, 193
trout, 159
turnips, 198,247

umbrella palm, 158
underground barriers, 161, 163

valleys, 16
vallisneria, 158
vegetables, 199
vegetation, 64
vetch, 134
vine maple, 133,134
vineyard, 282
Virginia creeper, 134
vitamins, 298

water, 14,139
water flea, 159
water heaters, 128-132
water lettuce, 158
water lily, 158
watermelons, 198,248
water pennywort, 17,27
water plants, 158
water poppy, 158
water pressure, 141
watershed, 15,140,160

INDEX

water tanks, 141-146
water wheels, 117-120
watering, 176,250,296,297
weeds, 65,271
wheat, 198,249
wild rice, 158,236
wild rye, 133
willow, 17,33,133
wind, 61,121,270
windbreaks, 79,270
windchargers, 124
windmills, 121

windows, 102
woodfern, crested, 17,21
wood heaters, 130
woodstoves, 131
woodwardia fern, 17,34

yellowjackets, 279

zinc, 257-259
zodiac, 190
zooplankton, 159,256

NOTES

NOTES